PRINTED IN U.S.A.
Lithoprinted by Edwards Brothers, Inc., Lithoprinters
Ann Arbor, Michigan, 1941

PREFACE

The theory of operators in Hilbert space has its roots in the theory of orthogonal functions and integral equations. Its growth spans nearly half a century and includes investigations by Fredholm, Hilbert, Weyl, Hellinger, Toeplitz, Riesz, Frechet, von Neumann and Stone. While this subject appeals to the imagination, it is also satisfying because due to its present abstract methods, questions of necessity and sufficiency are satisfactorily handled. One can therefore be confident that its developement is far from complete and eagerly await its further growth.

These notes present a set of results which we may call the group germ of this theory. We concern ourselves with tho structure of a single normal operator and at the end present the reader with a reading guide which, we believe, will give him a clear and reasonably complete picture of the theory.

Fundamentally the treatment given here is based on the two papers of Professor J. von Neumann referred to at the end of Chapter I. An attempt however has been made to unify this treatment and also recast it in certain respects. (Cf. the introductory paragraphs of Chapter IX). The elementary portions of the subject were given as geometrical a form as possible and the integral representations of unitary, self-adjoint and normal operators were linked with the canonical resolution.

In presenting the course from which these notes were taken, the author had two purposes in mind. The first was to present the most elementary course possible on this subject. This seemed desirable since only in this way could one hope to reach the students of physics and of statistics to whom the subject can offer so much. The second purpose was to emphasize those notions which seem to be proper to linear spaces and in particular to Hilbert space and omitting other notions as far as possible. The importance of the combination of various notions cannot be over-emphasized but there is a considerable gain in clarity in first treating them separately. These purposes are not antagonistic. We may point out that the theoretical portions of this work, except §4 of Chapter III, can be read without a knowledge of Lebesgue integration.

On the other hand, for these very reasons, the present work cannot claim to have supplanted the well-known treatise of M. H. Stone or the lecture notes of J. von Neumann. It is simply hoped that the student will find it advantageous to read the present treatment first and follow the reading guides given in Chapters XI and XII in consulting Stone's treatise and the more recent literature.

To those familiar with the subject, it will hardly be necessary to point out that the influence of Professor von Neumann is effective throughout the present work. Professor Bochner of Princeton University has also taken a kind interest in this work and made a number of valuable suggestions. I am also deeply grateful to my brother, Mr. John E. Murray, whose valuable assistance in typing these lecture notes, was essential to their preparation.

Columbia University,
New York, N. Y.
May, 1940

F. J. Murray

TABLE OF CONTENTS

TABLE OF CONTENTS

TABLE OF CONTENTS

Page

CHAPTER I

The expressions:

$$T_1 f = \int_0^1 k(x,y)f(y)dy$$

or

$$T_2 f = p(x)i\frac{d}{dx}f(x)+q(x)f(x)$$

or in the case of a function of two variables,

$$T_3 f = \frac{\partial^2 f}{\partial x^2}+\frac{\partial^2 f}{\partial y^2}$$

are linear operators. Thus the first two, when applicable, take a function defined on the unit interval into another function on the same interval.

Now if we confine our attention to functions $f(x)$ continuous on the closed unit interval and with a continuous derivative, we know that such a function can be expressed in the form,

$$f(x) = \Sigma_{\alpha=-\infty}^{\infty} x_\alpha \exp(2\pi i\alpha x)$$

where $x_\alpha = \int_0^1 f(x)\exp(-2\pi i\alpha x)dx$. If $T_1 f$ is of the same sort,

$$T_1 f = \Sigma_{\alpha=-\infty}^{\infty} y_\alpha \exp(2\pi i\alpha x)$$

where $y_\alpha = \int_0^1 T_1 f \ \exp(-2\pi i\alpha x)dx =$

$$\Sigma_{\beta=-\infty}^{\infty} x_\beta \int_0^1 T_1(\exp(2\pi i\beta x))\exp(-2\pi i\alpha x)dx = \Sigma_{\beta=-\infty}^{\infty} x_\beta a_{\beta,\alpha}.$$

Now for T_3 a somewhat similar argument holds, although it is customary to use a double summation.

The important thing to notice is that the operator equation

$$Tf = g$$

can be, in these cases, replaced by an infinite system of linear equations in an infinite number of unknowns. We shall prove that this can be done in far more general circumstances.

One might attempt to solve such an infinite system of equations by substituting a finite system and then passing to the limit, for example one might take the first n equations and ignore all but the first n unknowns. But this process is ineffective in general and introduces certain particular difficulties of its own.

Other methods must be sought. The choice of the functions $\exp(2\pi i\alpha x)$ corresponds to a choice of a system of coordinate axes in the case of a finite number of unknowns. In the finite case for a symmetrical operator, the coordinate system can be chosen, so that,

$$a_{\alpha,\beta} = \lambda_\alpha \delta_{\alpha,\beta} \qquad \begin{array}{l} \delta_{\alpha,\beta} = 0 \quad \text{if } \alpha \neq \beta, \\ \delta_{\alpha,\alpha} = 1. \end{array}$$

(Cf. Chap.VII, §1, Lemma 4) Correspondingly in the infinite case we would seek a complete set of functions ϕ_α such that

$$T\phi_\alpha(x) = \lambda_\alpha \phi_\alpha(x).$$

When this has been done, inverting the equation becomes a simple process. For example, consider, $T = i\frac{d}{dx}$, $\phi_\alpha(x) = \exp(-2\pi i\alpha x)$. While this is in general impossible, nevertheless an effective method of generalizing the result in the finite dimensional case exists and we shall discuss it in the present work.

We shall want to give our discussion its most general form and for that reason we consider not the set of functions whose square is summable, but rather an abstract space which has just those properties of this set which are needed in our developement. This space, $\mathfrak{H}$, is called Hilbert space and we shall show in Chapter II the existence of something equivalent to orthogonal sets of functions.

In Chapter III, we discuss L_2 and other realizations of abstract Hilbert space. In Chapters IV, V and VI, linear transformations are studied and certain preliminary properties established. Also a notion "weak convergence," which is of considerable interest in the theory of abstract spaces, is introduced to establish Theorem V of Chapter V.

In Chapter VII, we shall develope the needed generalization of the notion of an operator in diagonal form. In Chapter IX, we shall show that a self-adjoint operator even if it is discontinuous, is expressable in the diagonal form.

Symmetry is not sufficient in the discontinuous case as we shall see. The distinction between symmetric and self-adjoint transformations is brought out in Chapter X. In Chapter XI, a brief outline of further developements in the theory is given.

Except for Chapter XI, our discussion is based on the following:

(1) F. Riesz and E. R. Lorch. Trans. of the Amer. Math. Soc. Vol. 39, pp. 331-340 (1936).

(2) M. H. Stone. Colloquium Publications of Amer. Math. Soc. Vol. XV (1932).

(3) J. von Neumann. Math. Annalen Bd. 102 pp. 49-131 (1929).

(4) J. von Neumann. Annals of Mathematics, 2nd series, Vol. 33, pp. 294-310 (1932).

CHAPTER II

The axiomatic treatment of Hilbert space was first given by J. von Neumann in (4) pp. 64-69. He proposed the definition given below. We follow here the discussion given by Stone (2). (Numerals in parentheses refer to the references cited at the end of Chapter I.)

DEFINITION 1.1. A class $\mathfrak{H}$ of elements f, g, ... is called a Hilbert space if it satisfies the following postulates:

POSTULATE A. $\mathfrak{H}$ is a linear space; that is,

(1) there exists a commutative and associative operation denoted by $+$, applicable to every pair f, g of elements of $\mathfrak{H}$, with the property that $f+g$ is also an element of $\mathfrak{H}$.

(2) there exists a distributive and associative operation, denoted by $\cdot$, applicable to every pair (a,f), where a is a complex number and f is an element of $\mathfrak{H}$;

(3) in $\mathfrak{H}$ there exists a null element denoted by θ with the properties

$$f+\theta = f, \qquad a\cdot\theta = \theta, \qquad 0\cdot f = \theta.$$

POSTULATE B. There exists a numerically-valued function (f,g) defined for every pair f, g of elements of $\mathfrak{H}$, with the properties:

(1) $(af,g) = a(f,g)$.

(2) $(f_1+f_2,g) = (f_1,g)+(f_2,g)$.

(3) $(g,f) = \overline{(f,g)}$.

(4) $(f,f) \geq 0$.

(5) $(f,f) = 0$ if and only if $f = \theta$.

The not-negative real number $(f,f)^{1/2}$ will be denoted for convenience by $|f|$.

POSTULATE C. For every n, $n = 1, 2, 3, \ldots$, there exists a set of n linearly independent elements of $\mathfrak{H}$; that is, elements $f_1, \ldots, f_n$ such that the equation $a_1f_1 + \ldots + a_nf_n = \theta$ is true only when $a_1 = \ldots = a_n = 0$.

POSTULATE D. $\mathfrak{H}$ is separable; that is, there exists a denumerably infinite set of elements of $\mathfrak{H}$, $f_1, f_2, \ldots$, such that for every g in $\mathfrak{H}$ and every positive ϵ there exists an $n = n(g, \epsilon)$ for which $|f_n - g| < \epsilon$.

POSTULATE E. $\mathfrak{H}$ is complete; that is, if a sequence $\{f_n\}$ of elements of $\mathfrak{H}$ satisfies the condition

$$|f_m - f_n| \longrightarrow 0, \quad m, n \longrightarrow \infty$$

then there exists an element f of $\mathfrak{H}$ such that

$$|f - f_n| \longrightarrow 0, \quad n \longrightarrow \infty.$$

In this statement of the postulates, certain notations and conventions were introduced. These are (a) $-f = (-1)f$, (b) $f - g = f + (-1)g$, (c) $\bar{a}$ denotes the complex conjugate of a. We shall also use (d) $af = a \cdot f$, (e) $R(a)$ is the real part of the complex number a, $J(a)$ is the imaginary part of a, (f) $|a|$ is the absolute value of a.

The properties B(1) - B(5) imply

B(6) $(f, ag) = \bar{a}(f, g)$
B(7) $(f, g_1 + g_2) = (f, g_1) + (f, g_2)$
B(8) $|af| = |a| \cdot |f|$

We shall next prove that these imply

B(9) $|(f,g)| \leq |f| \cdot |g|$, the equality sign holding if and only if f and g are linearly dependent.

For if η, λ and μ are real and $\lambda^2 + \mu^2 \neq 0$ then

$$\lambda f + \exp(i\eta)\mu g$$

is θ for some choice of η, λ and μ if and only if f and g are linearly independent. Thus by B(4) and B(5)

$$(\lambda f+\exp(i\eta)\mu g,\lambda f+\exp(i\eta)\mu g) \geq 0$$

and equality can only occur when f and g are linearly dependent.

Expanding by means of B(2), B(1), B(7) and B(6) and using, with B(3) the fact that for any complex a, $a+\bar{a} = 2R(a)$, we obtain

$$\lambda^2|f|^2+2\lambda\mu R(\exp(-i\eta)(f,g))+\mu^2|g|^2 \geq 0$$

with equality possible only if f and g are linearly dependent.

Now we can choose η so that $\exp(-i\eta)(f,g) = -|(f,g)|$. Then the equation becomes in the linearly independent case

$$\lambda^2|f|^2-2\lambda\mu|(f,g)|+\mu^2|g|^2 > 0.$$

Now for linear independence, $|g| \neq 0$ and thus if we let $\lambda = |g|$, $\mu = |(f,g)|/|g|$, we get

$$|f|^2\cdot|g|^2-|(f,g)|^2 > 0.$$

On the other hand if f and g are linearly dependent it is easily seen that the equality holds.

B(10) $|f+g| \leq |f|+|g|$, with equality possible only if f and g are linearly dependent.

Proof:

$$|f+g|^2 = (f+g,f+g) = |f|^2+2R(f,g)+|g|^2 \leq$$
$$|f|^2+2|f|\cdot|g|+|g|^2 = (|f|+|g|)^2$$

§2

A weaker restriction than B is the postulate:

POSTULATE B'. There exists a real valued function $|f|$ of elements of $\mathfrak{H}$ with the properties B(4), B(5), B(8), B(10).

The function $|f|$ is called the norm. If a space satisfies

postulates A, B', and E it is usually referred to as a Banach space.* If D is also satisfied, the space is called separable. Thus Hilbert space is a separable Banach space but there are, as we shall see, separable Banach spaces, which are not Hilbert spaces.

The relation between Hilbert space and a separable Banach space is clearer if we consider

$$\text{B}(11) \quad |f+g|^2+|f-g|^2 = 2(|f|^2+|g|^2)$$

This equation is an immediate consequence of the equation

$$|f \pm g|^2 = |f|^2 \pm 2R(f,g)+|g|^2$$

Thus in Hilbert space, we have B' and B(11) and it can be shown that B' and B(11) are sufficient to insure that a separable Banach space is a Hilbert space.**

The major purpose of this book is to give as simply as possible certain results in the theory of Hilbert space and these specific results do not hold in general separable Banach space. However the Hilbert space theory can be more clearly understood if one appreciates the precise dependence of this theory upon certain specific properties of Hilbert space. For this reason, we shall endeavor to give the fundamentals of our subject, without restricting ourselves to Hilbert space, to the largest extent consistent with our purpose.

§3

If the linear space $\mathcal{E}$ has a norm $|f|$, then $d(f,g) = |f-g|$ is a metric for the space, i.e., satisfies the conditions

(i) $d(f,g) \geqq 0$, $d(f,g) = 0$ if and only if $f = g$.

(ii) $d(f,g) = d(g,f)$.

(iii) $d(f,g) \leqq d(f,h)+d(h,g)$.

These conditions are consequences of B(4), B(5), B(8) and B(10).

* These spaces have been investigated in a famous treatise "Theorie des Operations Lineaires." by S. Banach (Warsaw (1932)).

** J. von Neumann and Jordan, Annals of Mathematics, vol.36 (1935), pp. 719-724.

Thus we are invited to introduce the notion of continuity in such a space.

DEFINITION 1. Let $F(f)$ be a function defined on a subset of $\mathcal{E}$. This subset is called the domain of F. Let f_0 be an element of the domain of F. If for every $\epsilon > 0$, it is possible to find a δ such that if f is in the domain of F and $|f-f_0| < \delta$, then $|F(f)-F(f_0)| < \epsilon$, we say that F is continuous at f_0. If F is continuous at every point of its domain, F is said to be continuous.

If F assumes only complex numbers as its values, it is called a functional. Thus $|f|$ itself is a continuous functional.

DEFINITION 2. A function $F(f)$ will be called additive if whenever f and g are in its domain, $af+bg$ is also in the domain for any two complex numbers a and b and $F(af+bg) = aF(f)+bF(g)$.

Notice that these definitions apply not only to functionals but even to functions, which assume values in any linear space.

LEMMA 1. For an additive function $F(f)$, and any f_0 in its domain, the following statements are equivalent.

(a) F is continuous at f_0.

(b) F is continuous at θ.

(c) There exists a C such that $|F(f)| \leq C|f|$ for every f in the domain of F.

We note firstly that if f_0 is in the domain of F, $f_0-f_0 = \theta$ is also in the domain of F and $F(\theta) = F(f_0)-F(f_0) = \theta'$. ($\theta'$ is the null element for the space of the values of F.)

The element f is in the domain of F if and only if $h = f-f_0$ is in the domain. Also

$$|f-f_0| = |h| = |h-\theta|$$

and

$$|F(f)-F(f_0)| = |F(f-f_0)| = |F(h)-\theta'| = |F(h)-F(\theta)|.$$

These statements give precisely the equivalence of (a) and (b) by a substitution.

We will show that (b) and (c) are equivalent. Suppose (b). Then if ϵ is taken as 1, the continuity at θ implies that there is a δ such that when $|h| < \delta$, h in the domain of F, then, $|F(h)| < 1$. Now if f is any arbitrary element of the domain, af (= af+1·θ) is in the domain for every a. Let $a = \delta/2\cdot|f|$. then $h = af$ is such that $|h| = |(\delta/2\cdot|f|)\cdot f| = \delta/2 < \delta$. Hence

$$1 > |F(h)| = |aF(f)| = |a||F(f)| = (\delta/2|f|)\cdot|F(f)|$$

or $|f|(2/\delta) > |F(f)|$, and hence $2/\delta$ is a constant for which (c) holds. Thus (b) implies (c).

Now let us suppose (c) and that an $\epsilon > 0$ has been given. Lot δ be such that $C\delta < \epsilon$. Then if $|h-\theta| < \delta$, we have $|F(h)-F(\theta)| = |F(h)| \leq C|h| < C\delta < \epsilon$. Thus F is continuous at θ and (c) implies (b).

THEOREM I. An additive function F(f) is continuous at every point if it is continuous at one point. An additive function F(f) is continuous if and only if there exists a C such that for every f in its domain F, $|F(f)| \leq C|f|$.

A set S in $\mathfrak{H}$ will be called additive if whenever f and g are in it, af+bg is in it for every pair of complex numbers a and b. A closed additive set, $\mathfrak{M}$, will be called a linear manifold.

It is easily verified that the closure of an additive set is also additive and hence is a linear manifold. This depends on the fact that the limit of a linear combination in a Banach space is the linear combination of the limits. Let $f_n \longrightarrow f$ and $g_n \longrightarrow g$. Then since

$$|af+bg-(af_n+bg_n)| = |a(f-f_n)+b(g-g_n)| \leq |a||f-f_n|+|b||g-g_n| \longrightarrow 0$$

we have that if f and g are in the closure of a linear set

af+bg is also.

THEOREM II. The domain of an additive function $F(f)$ is additive. If $F(f)$ is also continuous, with values in a complete space, there exists a continuous additive function, $[F]$ with the properties

(a) The domain of $[F]$ is the closure of the domain of F.

(b) If f is in the domain of F, $[F](f) = F(f)$.

This $[F]$ is unique.

The first statement is obvious from the definitions. We will prove our statements concerning $[F]$ by specifying its values uniquely. Now if f is in the domain of F, $[F](f) = F(f)$. Let f be any point of the closure of the domain of F. If $\{f_n\}$ is a sequence of elements of the domain of F, such that $f_n \longrightarrow f$, then the $F(f_n)$'s are also convergent since $|F(f_n)-F(f_m)| = |F(f_n-f_m)| \leqq C|f_n-f_m| \longrightarrow 0$ as m and $n \longrightarrow \infty$ Since the values of $F(f_n)$ are in a complete space, they must converge to an f^*. Any two sequences $\{f_n'\}$ and $\{f_n''\}$ with the same limit f must have $\lim F(f_n') = \lim F(f_n'')$ since otherwise the sequence of $F(f)$'s consisting of elements which are taken alternately from one and then the other sequence of $F(f)$'s would have no limit. Thus f^* depends only on f. We may take $[F](f) = f^*$. (No contradiction with the previous definition of $[F]$ on the domain of F is possible, for if f is in the domain of F, we may take $f_n = f$). Furthermore if $[F]$ is continuous, this must be the definition. Thus the conditions (a) and (b) determine $[F]$ precisely.

To complete our proof it is only necessary to show that $[F]$ is additive and continuous. The additivity is a consequence of the facts given in the paragraph preceding the theorem, that the closure of an additive set is a linear manifold and that the limit of a linear combination is the linear combination of the limits. The continuity is shown by noting that if C is such that $|F(f)| \leqq C|f|$ for every f in the domain of F, then $|[F](f)| \leqq C|f|$ for every f in the closure of this domain. Such C 's exist by Theorem I and the same theorem shows that this implies continuity.

§4

An additive functional which is defined for all $f \in \mathcal{E}$ and which is continuous, is called a linear functional. For a Banach space, $\mathcal{E}$, the set of linear functionals, $\mathcal{E}^*$, is again a Banach space as one can see as follows.

Firstly, we notice that the set of linear functionals satisfies Postulate A if we define the sum of two linear functionals $F+G$ by the equation

$$(F+G)(f) = F(f)+G(g)$$

and scalar multiplication by the equation

$$(aF)(f) = aF(f).$$

To prove Postulate B', we define $|F|$ as the gr. l. b. of the C's for which $|F(f)| \leqq C|f|$ for all $f \in \mathcal{E}$. $|F|$ is readily seen to be the least such C. B(4) and B(5) are obvious from this definition, B(8) and B(10) follow from the definition of scalar multiplication and of addition given in the preceding paragraph.

To prove Postulate E, we consider any sequence $\{F_n\}$ of linear functionals, and such that $|F_n-F_m| \longrightarrow 0$ as n and m $\longrightarrow \infty$. It is readily seen that for each element f of $\mathcal{E}$,

$$|F_m(f)-F_n(f)| = |(F_m-F_n)(f)| \leqq |F_n-F_m|\cdot|f| \longrightarrow 0$$

as n and $m \longrightarrow \infty$ and furthermore, this approach to zero is uniform on those f's for which $|f| = 1$. Thus $F_n(f)$ has a limit $F(f)$ for every f in the space. It is easily seen that $F(f)$ is additive and that there is a C such that $|F(f)| \leqq C\cdot|f|$ for every $f \in \mathcal{E}$.

Now given ϵ, take N so large that for n and $m > N$, $|F_n-F_m| < \epsilon$. This means that we have

$$|F_n(f)-F_m(f)| \leqq \epsilon|f|.$$

Let us fix m, and let $n \longrightarrow \infty$. We then obtain

$$|F(f)-F_m(f)| \leqq \epsilon|f|.$$

Thus for $m > N$, $|F-F_m| < \epsilon$. This implies that F is such that $F_m \longrightarrow F$ as $m \longrightarrow \infty$ and hence that $\mathcal{E}^*$ is complete.

THEOREM III. The set $\mathcal{E}^*$ of linear functionals on a Banach space $\mathcal{E}$ is again a Banach space.*

Now one of the essential facts concerning Hilbert space is that $\mathfrak{H}^*$ is equivalent to $\mathfrak{H}$. The specific relation is given by the following theorems.

THEOREM IV. If F is a linear functional defined on the Hilbert space $\mathfrak{H}$, then there exists a $g \in \mathfrak{H}$ such that for every $f \in \mathfrak{H}$, $F(f) = (f,g)$.

Proof: If $F = 0$, we can let $g = \theta$. Suppose then that $|F| > 0$. If we are given a sequence of positive numbers $\{\epsilon_n\}$ with $\epsilon_n \longrightarrow 0$, we can find a sequence $\{g_n'\}$ of elements such that

$$|F|\cdot|g_n'| \geqq |F(g_n')| \geqq (1-\epsilon_n)|F|\cdot|g_n'|$$

and $F(g_n') \neq 0$. If we multiply g_n' by $1/|F(g_n')|$ we obtain a sequence g_n with $F(g_n) = 1$ and

$$|F|\cdot|g_n| \geqq 1 \geqq (1-\epsilon_n)\cdot|F|\cdot|g_n|.$$

Now consider $|g_n+g_m|$. We have

$$|F|\cdot|g_n+g_m| \geqq |F(g_n+g_m)| = 2 \geqq |F|\cdot(1-\epsilon_n)\cdot|g_n|+|F|(1-\epsilon_m)\cdot|g_m|$$

or

$$|g_n+g_m| \geqq (1-\epsilon_n)\cdot|g_n|+(1-\epsilon_m)\cdot|g_m|.$$

Thus

$$|g_n-g_m|^2 = 2(|g_n|^2+|g_m|^2)-|g_n+g_m|^2 \leqq$$
$$2(|g_n|^2+|g_m|^2)-((1-\epsilon_n)\cdot|g_n|+(1-\epsilon_m)\cdot|g_m|)^2$$

---

* A proof of the fact that Postulate C for $\mathcal{E}$ implies C for $\mathcal{E}^*$ can readily be given if the Hahn-Banach Extension Theorem is shown (Cf. Banach loc. cit. pp. 27-29). This has the consequence that if F is additive and continuous on a linear subset G_1 its definition can be extended throughout the space without increasing the norm. A proof of this is not on the main line of our developement but if this is assumed, one would proceed as follows.

Let $f_1, \dots, f_n$, be n linearly independent elements of and G the set of linear combinations of these. It is easily seen that one can define n linearly independent linear functionals $F_1, \dots, F_n$ on G. These can then be extended to the whole space by the extension theorem and this does not effect their linear independence.

Since $|g_n| \longrightarrow 1/|F|$ and $\epsilon_n \longrightarrow 0$ we have $|g_n - g_m|^2 \longrightarrow 0$ as n and $m \longrightarrow \infty$. Hence the g_n's form a convergent sequence. Define g so that $g_n \longrightarrow g$. Then $|g| = 1/|F|$, $F(g) = 1$. Now if h is such that $F(h) = 0$, we have that

$$1 = |F(g)| = |F(g+\lambda h)| \leqq |F| \cdot |g+\lambda h| = |g+\lambda h|/|g|$$

or for every λ,

$$|g| \leqq |g+\lambda h|.$$

Squaring we must have

$$|g|^2 \leqq |g+\lambda h|^2 = |g|^2 + 2R(\lambda(h,g)) + |\lambda|^2 \cdot |h|^2.$$

Now we can choose λ so that $2R(\lambda(h,g)) = -2|\lambda| \cdot |(h,g)|$. Thus $|g|^2 \leqq |g|^2 - 2\eta \cdot |(h,g)| + \eta^2 |h|^2$ for every $\eta > 0$. But this is possible only if $|(h,g)| = 0$. Thus if $F(h) = 0$, $(h,g) = 0$.

If h is arbitrary, $h = F(h)g + h'$ where $F(h') = F(h - F(h)g) = F(h) - F(h) \cdot F(g) = 0$. Let $g_0 = (1/|g|^2)g$. Then

$$(h,g_0) = (F(h)g+h', g_0) = F(h)(g,g_0) + (h',g_0) = F(h)(1/|g|^2)(g,g) + (1/|g|^2) \cdot (h',g) = F(h),$$

using the fact that $(h',g) = 0$ since $F(h') = 0$. Thus g_0 satisfies the condition of the theorem.

The converse of Theorem IV is the following:

THEOREM V. The equation $(f,g) = F(f)$, $f \in \mathfrak{H}$ defines for each g, a linear functional F with $|F| = |g|$.

Proof: F is obviously additive. Also

$$|F(f)| = |(f,g)| \leqq |f| \cdot |g|.$$

This implies that F is continuous and $|F| \leqq |g|$. Since however $|F(g)| = |g|^2 = |g| \cdot |g|$, $|F| \geqq |g|$, and thus we obtain the theorem.

Theorem V tells us that (f,g) is continuous in each variable, separately. But since

$$\begin{aligned} |(f+\delta f, g+\delta g) - (f,g)| &= |(\delta f, g) + (f, \delta g) + (\delta f, \delta g)| \\ &\leqq |(\delta f, g)| + |(f, \delta g)| + |(\delta f, \delta g)| \\ &\leqq |\delta f| \cdot |g| + |f| \cdot |\delta g| + |\delta f| \cdot |\delta g|, \end{aligned}$$

it is easy to show that (f,g) is continuous in both variables.

§5

The relation between linear functionals and the elements of $\mathfrak{H}$ has the following consequences. Consider a set S in any Banach space $\mathcal{E}$. We can consider $S^{\perp\prime}$ the set of linear functionals F such that $F(f) = 0$ for every f in S. It can be shown without difficulty that $S^{\perp\prime}$ is a linear manifold in $\mathcal{E}^*$. (The additivity is obvious and the closure is shown, by recalling that if F is a limit of the sequence F_n, $F_n(g) \longrightarrow F(g)$ for every element in $\mathcal{E}$). If $\mathcal{E} = \mathfrak{H}$, we have corresponding to $S^{\perp\prime}$, a set $S^{\perp}$, in $\mathfrak{H}$, for which $F \in S^{\perp\prime}$ and $F(f) = (f,g)$ for all $f \in \mathfrak{H}$, imply $g \in S^{\perp}$. Thus ordinarily in a Banach space the orthogonal complement $S^{\perp\prime}$ to a set S must be regarded in $\mathcal{E}^*$, but in $\mathfrak{H}$ we may take $S^{\perp}$ in the space itself.

Now $S^{\perp}$ as we have defined it above consists of all the $g \in \mathfrak{H}$, for which $(f,g) = 0$ for all $g \in \mathfrak{H}$. This too is readily seen to be a linear manifold. In the case in which S is itself a linear manifold $\mathfrak{M}$, we have the essential theorem:

THEOREM VI. Let $\mathfrak{M}$ be a linear manifold in $\mathfrak{H}$ and let $\mathfrak{M}^{\perp}$ be as above. Then if f is an arbitrary element of $\mathfrak{H}$, $f = f_1+f_2$, where $f_1 \in \mathfrak{M}$, $f_2 \in \mathfrak{M}^{\perp}$, and this resolution is unique.

We first note that any such resolution $f = f_1+f_2$ must be unique since if we have $f = f_1+f_2$ and $f = f_1'+f_2'$, then $g = f_1-f_1' = f_2'-f_2$ is in both $\mathfrak{M}$ and $\mathfrak{M}^{\perp}$ and hence $|g|^2 = (g,g) = 0$ and $g = \theta$.

Now if f is in $\mathfrak{M}$, $f+\theta = f$ is the desired resolution. We can suppose then that f is not in $\mathfrak{M}$. Consider P the set of elements $f-g$, where $g \in \mathfrak{M}$. Let $r = \text{gr.l.b.}|f-g|$, $g \in \mathfrak{M}$. Now $r \neq 0$, since otherwise we will have a sequence g_n such that $|f-g_n| \longrightarrow 0$ and $g_n \longrightarrow f$. Since $\mathfrak{M}$ is closed this would imply that $f \in \mathfrak{M}$ contrary to our hypothesis.

We can therefore find a sequence h_n in the form $f-g_n$, $g_n \in \mathfrak{M}$ and such that $|h_n| \longrightarrow r$. Since $\frac{1}{2}(h_n+h_m) = f-\frac{1}{2}(g_n+g_m)$, we have

$$\tfrac{1}{2}|h_n+h_m| = |\tfrac{1}{2}(h_n+h_m)| \geqq r$$

or $|h_n+h_m| \geqq 2r$.

Then by B(11);

$$|h_n-h_m|^2 = 2(|h_n|^2+|h_m|^2)-|h_n+h_m|^2 \leqq 2(|h_n|^2+|h_m|^2)-4r^2.$$

Since $|h_n| \longrightarrow r$, we see that $|h_n-h_m| \longrightarrow 0$ as n and $m \longrightarrow \infty$ and thus the h_n's converge to some element h with $|h| = r$. The $g_n = f-h_n$ also converge to a $g \in \mathfrak{M}$ and thus $h = f-g$ is a minimal element of P.

Now if g' is any element of $\mathfrak{M}$, $h+\lambda g' = f-(g-\lambda g')$ is in the set P. Thus

$$|h| \leqq |h+\lambda g'|$$

for every value of λ. As in the proof of Theorem IV, this implies $(g',h) = 0$. Since g' was any arbitrary element of $\mathfrak{M}$, h must be in $\mathfrak{M}^{\perp}$.. Thus the resolution $f = g+h$ is the desired one with $g \in \mathfrak{M}$, $h \in \mathfrak{M}^{\perp}$.*

COROLLARY 1. If $\mathfrak{M}_1$ and $\mathfrak{M}_2$ are two linear manifolds with $\mathfrak{M}_1 \subset \mathfrak{M}_2$, then if $f \in \mathfrak{M}_2$, $f = f_1+f_2$ where $f_1 \in \mathfrak{M}_1$ and $f_2 \in \mathfrak{M}_2 \cdot \mathfrak{M}_1^{\perp}$.

We must show that in the resolution $f = f_1+f_2$, $f_1 \in \mathfrak{M}_1$, $f_2 \in \mathfrak{M}_1^{\perp}$, we have $f_2 \in \mathfrak{M}_2$. Since f and f_1 are in $\mathfrak{M}_2$, this is true.

As a consequence of this, we have,

COROLLARY 2. If $\mathfrak{M}_1 \subset \mathfrak{M}_2$ but $\mathfrak{M}_1 \neq \mathfrak{M}_2$, then $\mathfrak{M}_2 \subset$ $\mathfrak{M}_2^{\perp} \subset \mathfrak{M}_1^{\perp}$ but $\mathfrak{M}_1^{\perp} \neq \mathfrak{M}_2^{\perp}$.

THEOREM VII. If $\mathfrak{M}$ is a linear manifold, $(\mathfrak{M}^{\perp})^{\perp} = \mathfrak{M}$.

Proof: It is readily seen that $\mathfrak{M} \subset (\mathfrak{M}^{\perp})^{\perp}$. We must demon-

---

* In some spaces other than $\mathfrak{H}$, it is possible, given an $F \in \mathcal{E}^*$, to find a g for which $F(g) = |F|\cdot|g|$. Furthermore inequalities similar to B(11) hold in these spaces (Cf. J. A. Clarkson, Trans. of the Amer. Math. Soc. Vol. 40, pp. 396-414, (1936)). These inequalities imply that the correspondence is not additive. Furthermore it is possible to show that in these spaces (not $\mathfrak{H}$), there is not even a generalized equivalent of Theorem VI. For it can be shown that there exists linear manifolds $\mathfrak{M}_1$, for which no linear manifold $\mathfrak{M}_2$ exists such that $\mathfrak{M}_1 \cdot \mathfrak{M}_2 = \{\theta\}$ and $f = f_1+f_2$, $f_1 \in \mathfrak{M}_1$, $f_2 \in \mathfrak{M}_2$ for every $f \in \mathcal{E}$. (Cf. F. J. Murray, Trans. of the Amer. Math. Soc. Vol. 41, pp. 138-152, (1937)).

strate $(\mathfrak{M}^{\perp})^{\perp} \subset \mathfrak{M}$. Let $f \in (\mathfrak{M}^{\perp})^{\perp}$. We have $f = f_1+f_2$, $f_1 \in \mathfrak{M}$, $f_2 \in \mathfrak{M}^{\perp}$. Since $f_2 = f-f_1$ and $f \in (\mathfrak{M}^{\perp})^{\perp}$, $f_1 \in \mathfrak{M} \subset (\mathfrak{M}^{\perp})^{\perp}$, we have f_2 also $\in (\mathfrak{M}^{\perp})^{\perp}$. Hence $f_2 \in (\mathfrak{M}^{\perp})(\mathfrak{M}^{\perp})^{\perp} = \{\Theta\}$ and $f = f_1 \in \mathfrak{M}$. Thus $f \in (\mathfrak{M}^{\perp})^{\perp}$ implies $f \in \mathfrak{M}$ and this completes the proof.

§6

If S is an arbitrary set of elements, let $\mathfrak{U}(S)$ denote the set of linear combinations of the elements of S, i.e., the set of $a_1f_1+ \dots +a_nf_n$, $f_i \in S$. For this notion, the following properties are easily obtainable: $\mathfrak{U}(S)^{\perp} = S^{\perp}$. If $S_2 \subset S_1$, $\mathfrak{U}(S_2) \subset \mathfrak{U}(S_1)$. Also if $S_2 \subset \mathfrak{U}(S_1)$, $\mathfrak{U}(S_2) \subset \mathfrak{U}(S_1)$.

The closure of $\mathfrak{U}(S)$, we denote by $\mathfrak{M}(S)$. For this, again we have, $\mathfrak{M}(S)^{\perp} = \mathfrak{U}(S)^{\perp} = S^{\perp}$, and if $S_2 \subset \mathfrak{M}(S_1)$, $\mathfrak{M}(S_2) \subset \mathfrak{M}(S_1)$. It follows from these and Theorem VII that $\mathfrak{M}(S) = (\mathfrak{M}(S)^{\perp})^{\perp} = (S^{\perp})^{\perp}$.

To develope these notions further, we prove the following lemma:

LEMMA 1. If $\mathcal{E}$ is a separable metric space, and S is a non-empty subset of $\mathcal{E}$, then there exists a finite or denumerably infinite subset S_1 of S which is dense in S.

Proof: Let $f_1, f_2, \dots$ be dense in $\mathcal{E}$. We define for each α, a set of elements $g_{\alpha,\beta}$ in S, which will be finite or denumerably infinite depending on α. Let r_α = gr.l.b. $|f_\alpha-g|$, $g \in S$. If $r_\alpha = 0$, we can chose a sequence $g_{\alpha,n}$ such that $g_{\alpha,n} \longrightarrow f_\alpha$. If r_α is not 0, we can find a $g \in S$ such that $|f-g| < 2r$. We let $g_{\alpha,1}$ be such a g and let $g_{\alpha,2}, g_{\alpha,3}, \dots$, remain undefined. In this last case $\frac{1}{2}|f_\alpha-g_\alpha| < r < |f_\alpha-g|$ for every $g \in S$. There is at most a denumerable number of the $g_{\alpha,\beta}$.

Now suppose a $g \in S$ and an $\epsilon > 0$ are given. Choose α so that $|f_\alpha-g| < \epsilon/3$. If $r_\alpha = 0$, we can find a $g_{\alpha,k}$ such that $|f_\alpha-g_{\alpha,k}| < 2\epsilon/3$. If $r_\alpha \neq 0$, we have that $|f_\alpha-g_{\alpha,1}| < 2|f_\alpha-g| < 2\epsilon/3$. Thus we can find a $g_{\alpha,k}$ such that $|g-g_{\alpha,k}| < \epsilon$. Thus the set $\{g_{\alpha,\beta}\}$ is dense.

A set of elements S, will be said to be orthonormal if for $\phi \in S$, $|\phi| = 1$ and if ϕ and $\psi \in S$ and $\phi \neq \psi$, then $(\phi,\psi)=0$.

THEOREM VIII. An orthonormal set S in $\mathfrak{H}$ can contain at most a denumerable set of elements.

Let $f_1, f_2, \ldots$ be a dense set in $\mathfrak{H}$. For each ϕ of S we choose an $\alpha = \alpha_\phi$ such that $|\phi - f_\alpha| < \frac{1}{2}$. This correspondence $\phi \sim \alpha$ is one-to-one. For by our choice of α, to each ϕ there is only one α. Furthermore to each α, there is at most one ϕ, since if ϕ and $\psi \in S$ are such that $\alpha = \alpha_\phi = \alpha_\psi$, then

$$|\phi-\psi| = |(\phi-f_\alpha)-(\psi-f_\alpha)| \leqq |(\phi-f_\alpha|+|\psi-f_\alpha| < 1.$$

But if $\phi \neq \psi$, then

$$|\phi-\psi|^2 = |\phi|^2 - 2R((\phi,\psi)) + |\psi|^2 = 2.$$

This contradicts $|\phi-\psi| < 1$ so we must have $\phi = \psi$.

Since the elements of S, are in a one-to-one correspondence with a subset of the positive integers, there is at most a denumerable number of them.

We can therefore enumerate the elements of an orthonormal set S, $\phi_1, \phi_2, \ldots$. Then the orthonormal condition can be written $(\phi_\alpha, \phi_\beta) = \delta_{\alpha,\beta}$ where $\delta_{\alpha,\beta}$ is the Kronecker symbol and equals zero if $\alpha \neq \beta$ and one for $\alpha = \beta$.

THEOREM IX. Given any denumerable set S not all of whose elements are θ, we can find an orthonormal set S_1 such that $\mathfrak{U}(S_1) = \mathfrak{U}(S)$.

Let $g_1, g_2, \ldots$ be the given set S. Let k_1 be the least integer such that $g_{k_1} \neq \theta$. Let $\phi_1 = (1/|g_{k_1}|)g_{k_1}$. Then $|\phi_1| = 1$. Let $g'_\alpha = g_\alpha - (g_\alpha, \phi_1)\phi_1$. For $\alpha < k_1$, $g_\alpha = \theta$ and hence $g'_\alpha = \theta$.

$$\begin{aligned} g'_{k_1} &= g_{k_1} - (g_{k_1}, \phi_1)\phi_1 = g_{k_1} - (g_{k_1}, (1/|g_{k_1}|)g_{k_1})(1/|g_{k_1}|)g_{k_1} \\ &= g_{k_1} - (g_{k_1}, g_{k_1})(1/|g_{k_1}|^2)g_{k_1} = \theta. \end{aligned}$$

Also

$$(g'_\alpha, \phi_1) = (g_\alpha - (g_\alpha, \phi_1)\phi_1, \phi_1) = 0$$

since $(\phi_1, \phi_1) = 1$.

Now suppose that we have by the repetition of this process, arrived at a definition of a sequence $g_1^{(s)}, g_2^{(s)}, \ldots$, a number k_s and an orthonormal set $\phi_1, \ldots, \phi_s$, with the following properties:

(1) $k_{s-1} < k_s$ and $g_\alpha^{(s)} = \theta$ for $\alpha \leqq k_s$.

(2) $(g_\alpha^{(s)},\phi_i) = 0$ for $\alpha = 1, 2, \ldots$, $i = 1, \ldots, s$.

Now if $g_\alpha^{(s)} = \theta$ for every α, we do not define ϕ_{s+p}, $p \geqq 1$. But if $g_\alpha^{(s)} \neq \theta$, the method of the preceding paragraph can be applied to determine k_{s+1}, a ϕ_{s+1} and a sequence $\{g_\alpha^{(s+1)}\}$, with $(g_\alpha^{(s+1)},\phi_{s+1}) = 0$ for every α. The sequence $\{g_\alpha^{(s+1)}\}$ and k_{s+1} have property (1), above.

We also have for $i = 1, \ldots, s$, that

$$(\phi_{s+1},\phi_i) = (g^{(s)}_{k_{s+1}},\phi_i) = 0.$$

Thus $\phi_1, \ldots, \phi_{s+1}$ is an orthonormal set. By our construction, we already have property (1) for $s+1$ and we have property (2) in the case of $i = s+1$. If $i \leqq s$,

$$\begin{aligned}(g_\alpha^{(s+1)},\phi_i) &= (g_\alpha^{(s)}-(g_\alpha^{(s)},\phi_{s+1})\phi_{s+1},\phi_i) \\ &= (g_\alpha^{(s)},\phi_i)-(g_\alpha^{(s)},\phi_{s+1})(\phi_{s+1},\phi_i) = 0\end{aligned}$$

since $(g_\alpha^{(s)},\phi_i) = 0$ by the hypothesis of the deduction and $(\phi_{s+1},\phi_i) = 0$ as above.

Thus the process either stops with some s with $g_\alpha^{(s)} = \theta$ for every α or it continues indefinitely. In any case, we have defined an orthonormal set, $\phi_1, \phi_2, \ldots$. We note that each ϕ_i is a linear combination of the g's.

Furthermore each g_α is a linear combination of the ϕ's. for either $g_\alpha = \theta$ or there is a least s such that $g_\alpha^{(s)} = \theta$. Then

$$\begin{aligned}g_\alpha &= g_\alpha'+(g_\alpha,\phi_1)\phi_1 \\ &= g_\alpha''+(g_\alpha',\phi_2)\phi_2+(g_\alpha,\phi_1)\phi_1 \\ &= g_\alpha^{(s)}+(g_\alpha^{(s-1)},\phi_{s-1})\phi_{s-1}+ \cdots +(g_\alpha,\phi_1)\phi_1.\end{aligned}$$

Let S_1 be the orthonormal set $\phi_1, \ldots, \phi_n, \ldots$. Each g_α is then in $\mathfrak{U}(S_1)$ hence $S \subset \mathfrak{U}(S_1)$ and hence $\mathfrak{U}(S) \subset \mathfrak{U}(S_1)$. Similarly $\mathfrak{U}(S_1) \subset \mathfrak{U}(S)$, since each ϕ_α is a linear combination of the g_α's.

This process of "orthonormalizing" the sequence $g_1, g_2, \ldots$ is usually referred to as the "Gram-Schmidt" process.

THEOREM X. Given a set S, having elements other

than θ, it is possible to find an orthonormal set S_1 in $\mathfrak{U}(S)$ such that $\mathfrak{M}(S_1) = \mathfrak{M}(S)$.

Proof: By a previous lemma, there is a denumerable set S' dense in S. Then $\mathfrak{U}(S')$ is dense in $\mathfrak{U}(S)$. We can find an orthonormal set S_1 such that $\mathfrak{U}(S_1) = \mathfrak{U}(S')$. Thus $\mathfrak{U}(S_1)$ is dense in $\mathfrak{U}(S)$. Thus the closure of $\mathfrak{U}(S_1)$ equals the closure of $\mathfrak{U}(S)$ or $\mathfrak{M}(S_1) = \mathfrak{M}(S)$.

COROLLARY. For every $\mathfrak{M}$, there is an orthonormal set S_1 such that $\mathfrak{M}(S_1) = \mathfrak{M}$.

Proof: In Theorem X, we can let $S = \mathfrak{M}$.

LEMMA 2. Suppose $S = \{\phi_1, \phi_2, \dots\}$ is an infinite orthonormal set. Then for a sequence of numbers a_1, $a_2, \dots$,

$$\Sigma_{\alpha=1}^{\infty} a_\alpha \phi_\alpha = \lim_{n\to\infty} \Sigma_{\alpha=1}^{n} a_\alpha \phi_\alpha$$

exists if and only if $\Sigma_{\alpha=1}^{\infty} |a_\alpha|^2 < \infty$. We also have

$$|\Sigma_{\alpha=1}^{\infty} a_\alpha \phi_\alpha|^2 = \Sigma_{\alpha=1}^{\infty} |a_\alpha|^2$$

when either limit exists.

Proof: We note that for $m \geqq n+1$,

$$\begin{aligned} |\Sigma_{\alpha=n+1}^{m} a_\alpha \phi_\alpha|^2 &= (\Sigma_{\alpha=n+1}^{m} a_\alpha \phi_\alpha, \Sigma_{\beta=n+1}^{m} a_\beta \phi_\beta) \\ &= \Sigma_{\alpha=n+1}^{m} \Sigma_{\beta=n+1}^{m} a_\alpha \bar{a}_\beta (\phi_\alpha, \phi_\beta) \\ &= \Sigma_{\alpha=n+1}^{m} \Sigma_{\beta=n+1}^{m} a_\alpha \bar{a}_\beta \delta_{\alpha,\beta} \\ &= \Sigma_{\alpha=n+1}^{m} |a_\alpha|^2. \end{aligned}$$

Thus if $f_n = \Sigma_{\alpha=1}^{n} a_\alpha \phi_\alpha$, $|f_n - f_m|^2 = \Sigma_{\alpha=n+1}^{m} |a_\alpha|^2$. Thus the convergence of the sequence f_n is equivalent to that of the partial sums $\Sigma_{\alpha=1}^{n} |a_\alpha|^2$. Furthermore putting $n = 0$ in our first equation we get

$$|\Sigma_{\alpha=1}^{\infty} a_\alpha \phi_\alpha|^2 = \lim_{m\to\infty} |\Sigma_{\alpha=1}^{m} a_\alpha \phi_\alpha|^2 = \lim_{m\to\infty} \Sigma_{\alpha=1}^{m} |a_\alpha|^2 = \Sigma_{\alpha=1} |a_\alpha|^2.$$

LEMMA 3. Suppose $\phi_1, \dots, \phi_n$ is a (finite) orthonormal set. For $f \in \mathfrak{H}$, we define $a_\alpha = (f, \phi_\alpha)$. Then

$f-\Sigma_{\alpha=1}^{n} a_\alpha\phi_\alpha$ is orthogonal to ϕ_α for $\alpha = 1, \ldots, n$, and $|f|^2 \geqq \Sigma_{\alpha=1}^{n} |a_\alpha|^2$.

Proof: For $\beta = 1, \ldots, n$,

$$(f-\Sigma_{\alpha=1}^{n} a_\alpha\phi_\alpha, \phi_\beta) = (f,\phi_\beta)-\Sigma_{\alpha=1}^{n} a_\alpha(\phi_\alpha,\phi_\beta)$$
$$= (f,\phi_\beta)-a_\alpha = 0.$$

Thus

$$0 \leqq |f-\Sigma_{\alpha=1}^{n} a_\alpha\phi_\alpha|^2 = (f-\Sigma_{\alpha=1}^{n} a_\alpha\phi_\alpha, f-\Sigma_{\beta=1}^{n} a_\beta\phi_\beta)$$
$$=((f-\Sigma_{\alpha=1}^{n} a_\alpha\phi_\alpha, f)-\Sigma_{\alpha=1}^{n} a_\alpha(\phi_\alpha, f-\Sigma_{\beta=1}^{n} a_\beta\phi_\beta)$$
$$= (f-\Sigma_{\alpha=1}^{n} a_\alpha\phi_\alpha, f) = (f,f)-\Sigma_{\alpha=1}^{n} a_\alpha(\phi_\alpha, f)$$
$$= |f|^2-\Sigma_{\alpha=1}^{n} a_\alpha\bar{a}_\alpha = |f|^2-\Sigma_{\alpha=1}^{n} |a_\alpha|^2.$$

This completes the proof of the lemma.

COROLLARY. If $\phi_1, \phi_2, \ldots$ is an infinite orthonormal set and if for a given $f \in \mathfrak{H}$, we define $a_\alpha = (f,\phi_\alpha)$, then $\Sigma_{\alpha=1}^{\infty} a_\alpha\phi_\alpha$ exists and $f-\Sigma_{\alpha=1}^{\infty} a_\alpha\phi_\alpha$ is orthogonal to ϕ_α for $\alpha = 1, 2, \ldots$ and $|f|^2 \geqq \Sigma_{\alpha=1}^{\infty} |a_\alpha|^2$.

Since $|f|^2 \geqq \Sigma_{\alpha=1}^{n} |a_\alpha|^2$, we must have $|f|^2 \geqq \Sigma_{\alpha=1}^{\infty} |a_\alpha|^2$. This implies that $\Sigma_{\alpha=1}^{\infty} a_\alpha\phi_\alpha$ exists by Lemma 2. Since

$$f-\Sigma_{\alpha=1}^{\infty} a_\alpha\phi_\alpha = \lim_{n\to\infty} (f-\Sigma_{\alpha=1}^{n} a_\alpha\phi_\alpha)$$

Lemma 3 implies $(f-\Sigma_{\alpha=1}^{\infty} a_\alpha\phi_\alpha, \phi_\beta) = 0$ for $\beta = 1, 2, \ldots$.

THEOREM XI. If $\mathfrak{M}$ is a closed linear manifold, we can find an orthonormal set S_1, $\phi_1, \phi_2, \ldots$ (finite or infinite) such that $\mathfrak{M}(S_1) = \mathfrak{M}$. For every $f \in \mathfrak{H}$ when we define $a_\alpha = (f,\phi_\alpha)$, we have that $|f|^2 \geqq \Sigma_\alpha |a_\alpha|^2$ and $\Sigma_\alpha a_\alpha\phi_\alpha = f_1$, exists and is in $\mathfrak{M}$. If we define f_2 as $f-f_1$, then $f_2 \in \mathfrak{M}^\perp$ and $f = f_1+f_2$ is the resolution of Theorem VI. $|f|^2 = |f_1|^2+|f_2|^2$, $f \in \mathfrak{M}$ if and only if $|f|^2 = \Sigma_\alpha |a_\alpha|^2$.

The first sentence is the corollary to Theorem X restated. The second sentence is a consequence of Lemma 3 and its corollary.

Since $S_1^\perp = \mathfrak{M}^\perp$, $f_2 \in \mathfrak{M}^\perp$ also follows from Lemma 3 and its corollary. $f = f_1 + f_2$ is the resolution of Theorem VI since that resolution is unique. Since f_1 and f_2 are orthogonal, $|f|^2 = |f_1|^2 + |f_2|^2$. By Lemma 2 (and its proof), we can show that $|f_1|^2 = \Sigma_\alpha |a_\alpha|^2$. Thus $|f|^2 = \Sigma_\alpha |a_\alpha|^2$ if and only if $|f_2|^2 = 0$ or $f = f_1 \in \mathfrak{M}$.

If Theorem XI is applied to $\mathfrak{H}$ as a closed linear manifold we obtain:

THEOREM XII. There exists an orthonormal set S_1, ϕ_1, ϕ_2, ... such that $\mathfrak{M}(S_1) = \mathfrak{H}$. To every $f \in \mathfrak{H}$ we can find a sequence a_1, a_2, ... , $a_\alpha = (f, \phi_\alpha)$ with $|f|^2 = \Sigma_{\alpha=1}^\infty |a_\alpha|^2$ and $f = \Sigma_{\alpha=1}^\infty a_\alpha \phi_\alpha$. If $f \sim \{a_1, a_2, \ldots\}$ and $g \sim \{b_1, b_2, \ldots\}$ then

$$(f,g) = \Sigma_{\alpha=1}^\infty a_\alpha \bar{b}_\alpha .$$

In connection with the last sentence, we note that

$$\begin{aligned}(f,g) &= (\Sigma_{\alpha=1}^\infty a_\alpha \phi_\alpha, \Sigma_{\beta=1}^\infty b_\beta \phi_\beta) \\ &= \lim_{n\to\infty} (\Sigma_{\alpha=1}^n a_\alpha \phi_\alpha, \Sigma_{\beta=1}^\infty b_\beta \phi_\beta) \\ &= \lim_{n\to\infty} \Sigma_{\alpha=1}^n a_\alpha (\phi_\alpha, \Sigma_{\beta=1}^\infty b_\beta \phi_\beta) \\ &= \lim_{n\to\infty} \Sigma_{\alpha=1}^n a_\alpha \lim_{m\to\infty} (\phi_\alpha, \Sigma_{\beta=1}^m b_\beta \phi_\beta) \\ &= \lim_{n\to\infty} \Sigma_{\alpha=1}^n a_\alpha \bar{b}_\alpha = \Sigma_{\alpha=1}^\infty a_\alpha \bar{b}_\alpha .\end{aligned}$$

The correspondence $f \sim \{a_1, a_2, \ldots\}$ has certain other obvious properties. For instance

$$f+g \sim \{a_1+b_1, a_2+b_2, \ldots\}$$

$$af \sim \{aa_1, aa_2, \ldots\}$$

$$\theta \sim \{0, 0, \ldots\}$$

$$\phi_\alpha \sim \{\delta_{\alpha,1}, \delta_{\alpha,2}, \ldots\}.$$

An orthonormal set S_1 such that $\mathfrak{M}(S_1) = \mathfrak{H}$ is called complete.

CHAPTER III

REALIZATIONS OF HILBERT SPACE

§1

In this chapter, we will give certain examples of $\mathfrak{H}$. Our method of procedure will be to specify a set of elements, f, define + and a· so that Postulate A is satisfied, then define (f,g) so as to yield Postulate B. Postulate C will be in general almost trivial and it will then be necessary to establish Postulates D and E.

Thus we will deal with the sets $\mathfrak{H}'$ which are known to satisfy Postulates A and B. As we remarked in Chapter II §3, $\mathfrak{H}'$ is a linear metric space. Thus we can apply to $\mathfrak{H}'$, certain known theorems on metric spaces directly and this will in general simplify the proofs of seperability. Thus if S_1 is dense in S_2, the closure of S_1 is the closure of S_2. Furthermore, the notions $\mathfrak{U}(S)$ and $\mathfrak{M}(S)$ can be defined and we have the lemma.

LEMMA 1. If $\mathfrak{H}'$ satisfies Postulates A and B and if S is a denumerable set in $\mathfrak{H}'$, then $\mathfrak{M}(S)$ is separable.

Proof. Since $\mathfrak{M}(S)$ is the closure of $\mathfrak{U}(S)$, if $\mathfrak{U}(S)$ is separable, $\mathfrak{M}(S)$ is separable.. But $\mathfrak{U}(S)$ consists of elements in the form $\Sigma_{\alpha=1}^{n} a_\alpha f_\alpha$, $f_\alpha \in S$. Now let $\epsilon > 0$ be given. Let r_α be a number in the form $\rho_1 + i\rho_2$ where ρ_1 and ρ_2 are rational numbers and such that $|f_\alpha| \cdot |a_\alpha - r_\alpha| < \epsilon/n$. Then $|\Sigma_{\alpha=1}^{n} a_\alpha f_\alpha - \Sigma_{\alpha=1}^{n} r_\alpha f_\alpha|$ is easily seen to be $< \epsilon$. Thus the set of elements $\Sigma_{\alpha=1}^{n} r_\alpha f_\alpha$ (which we will denote by $\mathfrak{U}_r(S)$) is dense in $\mathfrak{U}(S)$.

But $\mathfrak{U}_r(S)$ has only a denumerable number of elements. For let us enumerate the elements of S; $f_1, f_2, \ldots$. We then see that the set of elements $r_1 f_1 + \ldots + r_n f_n$ for n fixed must be denumerable since an n 'tuple sequence can be rearranged in a single sequence. Now $\mathfrak{U}_r(S)$ is the set of all of these, i.e., for every n, and hence is a denumerable sum of denumerable sets. Thus $\mathfrak{U}_r(S)$ is denumerable.

But $\mathfrak{U}_r(S)$ is dense in $\mathfrak{U}(S)$ whose closure is $\mathfrak{M}(S)$ and thus $\mathfrak{M}(S)$ must be separable.

To show the infinite dimensionality i.e., Postulate C, the following lemma is useful.

LEMMA 2. Let $\mathfrak{H}'$ satisfy Postulates A and B. Let $\phi_1, \dots, \phi_n$ be n non-zero elements of $\mathfrak{H}'$, with the property that $(\phi_i, \phi_j) = 0$, if $i \neq j$. Then $\phi_1, \dots, \phi_n$ are linearly independent

For if $a_1\phi_1 + \dots + a_n\phi_n = \theta$, we have $0 = (\theta, \phi_i) = (\Sigma_{\alpha=1}^{n} a_\alpha\phi_\alpha, \phi_i) = a_i(\phi_i, \phi_i)$. Since $(\phi_i, \phi_i) \neq 0$, we have $a_i = 0$. Thus $a_1\phi_1 + \dots + a_n\phi_n = \theta$ implies $a_1 = \dots = a_n = 0$ and the ϕ_i 's are linearly independent.

§2

DEFINITION 1. Let l_2 denote the set of sequence of complex numbers $\{a_1, a_2, \dots\}$ such that $\Sigma_{\alpha=1}^{\infty} |a_\alpha|^2 < \infty$. We define

$$\{a_1, a_2, \dots\} + \{b_1, b_2, \dots\} = \{a_1+b_1, a_2+b_2, \dots\}$$
$$a\{a_1, a_2, \dots\} = \{aa_1, aa_2, \dots\}$$
$$\theta = \{0, 0, \dots\}$$
$$(\{a_1, a_2, \dots\}, \{b_1, b_2, \dots\}) = \Sigma_{\alpha=1}^{\infty} a_\alpha \overline{b}_\alpha.$$

We note that $|a+b|^2 \leq 2(|a|^2 + |b|^2)$ and thus if $\{a_1, a_2, \dots\}$ and $\{b_1, b_2, \dots\}$ are in l_2, then $\{a_1+b_1, a_2+b_2, \dots\}$ is also. $\{aa_1, aa_2, \dots\}$ obviously is in l_2 if $\{a_1, a_2, \dots\}$ is. Now it can be shown in precisely the same way as we we established B(9) in §1, Chapter II, that

$$(\Sigma_{\alpha=1}^{n} |a_\alpha \overline{b}_\alpha|)^2 = (\Sigma_{\alpha=1}^{n} |a_\alpha| \cdot |b_\alpha|)^2 \leq (\Sigma_{\alpha=1}^{n} |a_\alpha|^2)(\Sigma_{\beta=1}^{n} |b_\beta|^2)$$

Thus if $\Sigma_{\alpha=1}^{\infty} |a_\alpha|^2 < \infty$ and $\Sigma_{\beta=1}^{\infty} |b_\beta|^2 < \infty$, then the sum $\Sigma_{\alpha=1}^{\infty} a_\alpha \overline{b}_\alpha$ is absolutely convergent and thus the inner product is defined for every pair of elements of l_2.*

---

* If we consider l_p, $p > 1$, the set of sequences for which $\Sigma_{\alpha=1}^{\infty} |a_\alpha|^p < \infty$, thee operation + and a· will be also universally applicable. But to form an inner product, we would take $\{a_1, a_2, \dots\}$ in l_p and $\{b_1, b_2, \dots\}$ in $l_{p'}$ where $1/p + 1/p' = 1$. (Cf. S. Banach. loc. cit. pp. 67-68) This corresponds to the fact that $l_p^* = l_{p'}$ for these spaces. (Cf. Chapter II, §4 above) However the completeness, separability and infinite dimensionality of l_p can be demonstrated in a manner quite analogous to the corresponding proofs for l_2.

THEOREM I. l_2 is a Hilbert space.

Proof. From the previous definitions, it is readily seen that Postulates A and B are satisfied.

To show Postulate C we notice that S_1, the sequence of elements $\phi_1, \phi_2, \ldots$ where $\phi_\alpha = \{\delta_{\alpha,1}, \delta_{\alpha,2}, \ldots\}$ is a denumerable orthonormal set. Thus given n, we take the first n of the ϕ_α's, and Lemma 2 of the preceding section tells us that these are linearly independent.

To show Postulate D, we note that $\mathfrak{U}(S_1)$ is the set of sequences $\{a_1, a_2, \ldots\}$ for which there is an N such that if $n \geqq N$, $a_n = 0$. Now $\mathfrak{U}(S_1)$ is dense in l_2 since if $f \sim \{a_1, a_2, \ldots\}$ and $\epsilon > 0$ is given, we can choose an N so that if $n \geqq N$, $\Sigma_{\alpha=1}^{n} |a_\alpha|^2 > |f|^2 - \epsilon^2$. Now if f_n is the sequence $\{a_1, \ldots, a_n, 0, \ldots\}$ then $f_n \in \mathfrak{U}(S_1)$ and $|f - f_n| < \epsilon$. Thus $l_2 = \mathfrak{M}(S_1)$ and Lemma 1 now implies the separability postulate.

To show the completeness, we must consider a sequence of elements $\{f_n\}$ such that $|f_n - f_m| \longrightarrow 0$. We must show the existence of a sequence $g = \{b_1, b_2, \ldots\}$ such that $\Sigma_{\alpha=1}^{\infty} |b_\alpha|^2 < \infty$ and $|f_n - g| \longrightarrow 0$.

Now if $f_n = \{a_{n,1}, a_{n,2}, \ldots\}$ we define $f_{n,p} = \{a_{n,1}, \ldots, a_{n,p}, 0, \ldots\}$. Then for $\eta > 0$. there is an $N = N(\eta)$ such that if n and $m > N$, then

$$\eta^2 > |f_n - f_m|^2 = \Sigma_{\alpha=1}^{\infty} |a_{n,\alpha} - a_{m,\alpha}|^2$$

$$\geqq \Sigma_{\alpha=1}^{p} |a_{n,\alpha} - a_{m,\alpha}|^2 = |f_{n,p} - f_{m,p}|^2 \qquad (\alpha')$$

or

$$\eta > |f_n - f_m| \geqq |f_{n,p} - f_{m,p}| \qquad (\alpha)$$

for every p. Thus if we fix p, since $|f_n - f_m| \longrightarrow 0$, $|f_{n,p} - f_{m,p}| \longrightarrow 0$ and the sequence $f_{n,p}$ must be convergent. This means that for $q \leqq p$, $a_{n,q} \longrightarrow b_q$. We remark that since p can be taken indefinitely large that b_q is defined for every q. Let $g_p = \{b_1, \ldots, b_p, 0, \ldots\}$. Obviously $f_{n,p} \longrightarrow g_p$ as $p \longrightarrow \infty$.

Furthermore for $N = N(\eta)$, we have that $f_{N,p} \longrightarrow f_N$ as $p \longrightarrow \infty$. For $|f_N - f_{N,p}|^2 = \Sigma_{\alpha=p+1}^{\infty} |a_{N,\alpha}|^2$. Thus given $\eta > 0$, we can find a $P(\eta)$ such that if $p \geqq P(\eta)$,

$$|f_N - f_{N,p}| < \eta \qquad (\beta)$$

Now let n and m be $\geqq N(\eta)$, $p \geqq P(\eta)$. Then by (α), (β), and (α) again, we have

$$\begin{aligned} |f_n - f_{m,p}| &= |f_n - f_N + f_N - f_{N,p} + f_{N,p} - f_{m,p}| \\ &\leqq |f_n - f_N| + |f_N - f_{N,p}| + |f_{N,p} - f_{m,p}| \\ &\leqq 3\eta. \end{aligned} \qquad (\gamma)$$

If we let $m \longrightarrow \infty$, then $f_{m,p} \longrightarrow g_p$ and we obtain that for $n \geqq N(\eta)$, $p \geqq P(\eta)$,

$$|f_n - g_p| \leqq 3\eta. \qquad (\gamma.1)$$

If in particular we let $n = N$, we have $|f_N - g_p| \leqq 3\eta$ for every $p \geqq P(\eta)$. This implies $|g_p| \leqq |f_N| + 3\eta$ for $p \geqq P(\eta)$. Thus

$$\Sigma_{\alpha=1}^{p} |b_\alpha|^2 < (|f_N| + 3\eta)^2.$$

We may let $p \longrightarrow \infty$ and obtain

$$\Sigma_{\alpha=1}^{\infty} |b_\alpha|^2 \leqq (|f_N| + 3\eta)^2 < \infty.$$

Thus we may let $g = \{b_1, b_2, \ldots\}$.

We observe that $g_p \longrightarrow g$. This and $(\gamma.1)$ imply

$$|f_n - g| \leqq 3\eta \qquad (\gamma.2)$$

for $n \geqq N(\eta)$. This implies that $f_n \longrightarrow g$. The existence of a g with this property indicates the completeness of l_2 and we have demonstrated Theorem I.

When we recall Theorem XII of Chapter I, we obtain,

THEOREM II. Every Hilbert space is equivalent to l_2, in the sense that to every f of $\mathfrak{H}$, there is an element $\{a_n\}$ of l_2, $f \sim \{a_n\}$ and this correspondence is one-to-one, and preserves the operations $+$, $a\cdot$, θ and (f,g).

A set of postulates is said to be categorical if for any two realizations, there exists a one-to-one correspondence which preserves the relations of the postulates. Since any two realizations are in such a relations with l_2, they must be related in this way to each other. Thus the axioms of Hilbert space are categorical.

Now if a linear manifold, $\mathfrak{M}$ is infinite dimensional, i.e., satisfies postulate C, it satisfies all the postulates, with the original definitions of $+$, $a\cdot$, θ and $(\ ,\ )$. Thus we obtain:

COROLLARY. Every infinite dimensional manifold in a Hilbert space is equivalent to the space itself. It is also equivalent to l_2.

§3

DEFINITION 1. Let $n = 2, 3, \ldots$ be a given integer and let us suppose that we have n Hilbert spaces, $\mathfrak{H}_1, \ldots, \mathfrak{H}_n$. Now consider the n 'tuples of elements $\{f_1, \ldots, f_n\}$ with $f_i \in \mathfrak{H}_i$ for $i = 1, \ldots, n$. Define

$$\{f_1, \ldots, f_n\}+\{g_1, \ldots, g_n\} = \{f_1+g_1, \ldots, f_n+g_n\}$$
$$a\{f_1, \ldots, f_n\} = \{af_1, \ldots, af_n\}$$
$$\theta = \{\theta_1, \ldots, \theta_n\}$$
$$(\{f_1, \ldots, f_n\}, \{g_1, \ldots, g_n\}) = (f_1, g_1)+ \ldots +(f_n, g_n)$$

We call this set of n 'tuples, $\mathfrak{H}_1 \oplus \ldots \oplus \mathfrak{H}_n$.*

THEOREM III. $\mathfrak{H}_1 \oplus \ldots \oplus \mathfrak{H}_n$ is a Hilbert space.

The proof of this theorem is most elementary.

DEFINITION 2. Let $\mathfrak{H}_1, \mathfrak{H}_2, \ldots$ be a sequence of Hilbert spaces. We consider the sequences $\{f_1, f_2, \ldots\}$ such that $f_\alpha \in \mathfrak{H}_\alpha$ and $\Sigma_{\alpha=1}^{\infty} |f_\alpha|^2 < \infty$. We define

$$\{f_1, f_2, \ldots\}+\{g_1, g_2, \ldots\} = \{f_1+g_1, f_2+g_2, \ldots\}$$
$$a\{f_1, f_2, \ldots\} = \{af_1, af_2, \ldots\}$$
$$\theta = \{\theta_1, \theta_2, \ldots\}$$
$$(\{f_1, f_2, \ldots\}, \{g_1, g_2, \ldots\}) = (f_1, g_1)+(f_2, g_2)+ \ldots .$$

* If $\mathfrak{H}_1, \ldots, \mathfrak{H}_n$ are Banach and not Hilbert spaces, we can make similar definitions of the sum and scalar multiplications of n 'tuples. For the norm, it is however sometimes more convenient to use the definition

$$|\{f_1, \ldots, f_n\}| = (|f_1|^p + \ldots + |f_n|^p)^{1/p}.$$

The proof that the operations $+$, $a\cdot$, and $(,)$ are universally defined is quite analogous to the discussion given in §2 for l_2. We proceed to the theorem:

THEOREM IV. $\mathfrak{H}_1 \oplus \mathfrak{H}_2 \oplus \ldots$ is a Hilbert space.

A proof of this theorem quite analogous to the proof of Theorem I of this chapter is possible. Another proof can be obtained if one considers for each α, a complete orthonormal set $\{\phi_{\alpha,n}\}$ for $\mathfrak{H}_\alpha$. Thus if we have a sequence $\{f_1, f_2, \ldots\}$ with $\Sigma_{\alpha=1}^\infty |f_\alpha|^2 < \infty$, we can find for each f_α a sequence $a_{\alpha,1}, a_{\alpha,2}, \ldots$ of numbers with $a_{\alpha,\beta} = (f_\alpha, \phi_{\alpha,\beta})$ such that $|f_\alpha|^2 = \Sigma_{\beta=1}^\infty |a_{\alpha,}|^2$. Then $\Sigma_{\alpha=1}^\infty \Sigma_{\beta=1}^\infty |a_{\alpha,\beta}|^2 < \infty$. Now it can easily be shown that any method of corresponding a double sequence to a single sequence, determines an equivalence between $\mathfrak{H}_1 \oplus \mathfrak{H}_2 \oplus \ldots$ and l_2.

§4

DEFINITION. Let E be a measurable set of finite non-zero measure in an n dimensional space. Let $\mathfrak{L}_2$ consist of the Lebesgue measurable functions $f(P)$ defined on E and such that $\int_E |f(P)|^2 dP < \infty$. However two functions are to be regarded as identical if they differ only on a set of measure zero. Sum and $a\cdot$ are defined in the usual way for functions, θ is the function which is zero. (except possibly on a set of measure zero). Finally

$$(f,g) = \int_E f(P)\overline{g}(P)dP.$$

If n and E are not specified, it will be understood that $n = 1$ and E is the set of x such that $0 \leqq x \leqq 1$.

The operation $a\cdot$ is obviously universally defined in $\mathfrak{L}_2$. Since $|a+b|^2 \leqq 2(|a|^2+|b|^2)$ for a and b complex is is readily seen that $+$ is also universally defined. To obtain the corresponding result for $(,)$, we introduce the notion $f_A(P)$ for positive A 's. $f_A(P) = A$ if $f(P) > A$, $f_A(P) = f(P)$ if $-A \leqq f(P) \leqq A$ and $f_A(P) = -A$ if $f(P) < -A$. For f_A and g_A, we obtain without difficulty as in the proof of B(9) in §1, Chapter I that

$$(\int_E |f_A|\cdot|\bar{g}_A|dP)^2 \leqq (\int_E |f_A|^2 dP)(\int_E |g_A|^2 dP)$$
$$\leqq (\int_E |f|^2 dP)(\int_E |g|^2 dP).$$

Due to a well known theorem on a monotonically increasing sequence of positive measurable functions, this implies

$$(\int_E |f|\cdot|\bar{g}|dP)^2 \leqq (\int_E |f|^2 dP)(\int_E |g|^2 dP).$$

Hence $f(P)\cdot\bar{g}(P)$ is a sumable function of P and thus $(\, , \,)$ is universally defined.

THEOREM V. $\mathfrak{L}_2$ is a Hilbert space.

It is possible to show Postulates A and B without difficulty.

To show Postulate C we observe that since E is of finite non-zero measure it is possible to find n mutually exclusive measurable sets, $E_1, \ldots E_n$ included in E and each of non-zero measure. Let $\chi_1(P)$ be the characteristic function for E_1, i.e., the function which assumes the valus 1 on E_1 and 0 elsewhere. Then $\chi_1, \ldots, \chi_n$ are a set of mutually orthogonal non-zero functions to which Lemma 2 of §1 may be applied. Thus Postulate C is satisfied.

The proof of Postulate D depends on Lemma 1 of §1. Let S_1 denote the set of characteristic functions of the measurable sets $F \subset E$. Since every function of $\mathfrak{L}_2$ can be approximated by step functions, it follows that $\mathfrak{M}(S_1) = \mathfrak{L}_2$. Let S_2 denote the subset of these, in which the F is an intersection of an open set G with E. It is well known that S_2 is dense in S_1. Thus $\mathfrak{M}(S_2) = \mathfrak{M}(S_1) = \mathfrak{L}_2$. Let S_3 denote the subset of S_2, in which G is the interior of an n-dimensional cube whose faces have the equation $x_i = \rho_i$, where ρ_i is a rational number. Now any open set is a denumerable sum of such n-dimensional cubes regarded however as closed point sets, but whose interiors are mutually exclusive. The faces of the cubes in such a sum form a set of measure zero and thus it is possible to show that $\mathfrak{M}(S_3)$ contains S_2 and thus $\mathfrak{M}(S_3) = \mathfrak{M}(S_2) = \mathfrak{L}_2$. Lemma 1 of §1 of this chapter now implies Postulate D since S_3 is denumerable.

It remains to prove Postulate E for $\mathfrak{L}_2$. Let $f_1, f_2, \ldots$ be a sequence of elements such that $|f_n - f_m| \longrightarrow$ as n and

$m \longrightarrow \infty$. Let $\epsilon_1, \epsilon_2, \ldots$ be a sequence of positive numbers such that $\Sigma_{\alpha=1}^{\infty} \epsilon_\alpha < \infty$ and $\epsilon_\alpha > 0$. We take an increasing sequence of positive integers $n_1, n_2, \ldots$, such that if n and m are $> n_\alpha$, $|f_n - f_m| < \epsilon_\alpha$. It follows that $|f_{n_{\alpha+1}} - f_{n_\alpha}| < \epsilon_\alpha$. Thus

$$\Sigma_{\alpha=1}^{\infty} |f_{n_{\alpha+1}} - f_{n_\alpha}| < \Sigma_{\alpha=1}^{\infty} \epsilon_\alpha < \infty.$$

We let $k = \Sigma_{\alpha=1}^{\infty} |f_{n_{\alpha+1}} - f_{n_\alpha}|$.

Let

$$h_n(P) = \Sigma_{\alpha=1}^{n} |f_{n_{\alpha+1}}(P) - f_{n_\alpha}(P)|.$$

We have seen that

$$\int_E h_n(P)dP = \Sigma_{\alpha=1}^{n} \int_E |f_{n_{\alpha+1}}(P) - f_{n_\alpha}(P)|dP$$
$$\leq \Sigma_{\alpha=1}^{n} (m(E))^{1/2} |f_{n_{\alpha+1}} - f_{n_\alpha}| \leq m(E)^{1/2} \cdot k.$$

Thus the $h_n(P)$ are a monotomically increasing sequence of positive functions whose integrals are bounded. It follows that for almost every P, $h_n(P) \longrightarrow h(P) < \infty$.

We will next show that $h(P)$ is in $\mathfrak{L}_2$. For consider $h_A(P)$ (Cf. above). Obviously $h_{n,A}(P) \longrightarrow h_A(P)$ for almost every P, Since these functions are uniformly bounded, we have that

$$\int_E h_A^2(P)dP = \lim_{n\to\infty} \int_E h_{n,A}^2(P)dP$$
$$\leq \lim_{n\to\infty} \int_E h_n^2(P)dP = \lim_{n\to\infty} |h_n|^2$$
$$\leq \lim_{n\to\infty} (\Sigma_{\alpha=1}^{n} |f_{n_{\alpha+1}} - f_{n_\alpha}|)^2 = k^2.$$

Since this holds for every A, we have again an increasing sequence of positive functions $h_A^2(P)$ whose integral is bounded. Thus for the limit we have $\int_E h^2(P)dP \leq k^2 < \infty$.

From the existence of $h(P)$, we can conclude that the series $f_{n_1}(P) + (f_{n_2}(P) - f_{n_1}(P)) + (f_{n_3}(P) - f_{n_2}(P)) + \ldots$ is absolutely convergent for almost every P. Call the sum $g(P)$. Since $|g(P)| \leq |f_{n_1}(P)| + h(P)$ for almost every P, it is easily seen that $g(P)$ is in $\mathfrak{L}_2$.

It is readily seen that for almost every P

$$|g(P) - f_{n_\alpha}(P)|^2 = |\Sigma_{\beta=\alpha}^{\infty} (f_{n_{\beta+1}}(P) - f_{n_\beta}(P))|^2$$
$$\leq (\Sigma_{\beta=\alpha}^{\infty} |f_{n_{\beta+1}}(P) - f_{n_\beta}(P)|)^2 \leq h^2(P),$$

also $f_{n_\alpha}(P) \longrightarrow g(P)$. Thus by the majorant theorem of Lebesgue $\int_E |f_{n_\alpha}(P)-g(P)|^2 dP \longrightarrow 0$. Thus $f_{n_\alpha} \longrightarrow g$.

Suppose now that $\epsilon > 0$ is given. From the above we can find an α such that if $\beta \geqq \alpha$, $|g-f_{n_\beta}| < \epsilon/2$. We can find an $N = N(\epsilon/2)$ such that if n and $m \geqq N$, $|f_n-f_m| < \epsilon/2$. Let $P(\epsilon) = \max (N(\epsilon/2), n_\alpha)$. If $n \geqq P(\epsilon)$ and $n_\beta \geqq P(\epsilon)$ we have $|f_n-g| = |f_n-f_{n_\beta}+f_{n_\beta}-g| \leqq |f_n-f_{n_\beta}|+|f_{n_\beta}-g| < \epsilon$. Thus $f_n \longrightarrow g \in \mathfrak{L}_2$ and the completeness is shown.

If E is a measurable set of infinite measure, we can combine Theorem IV of §3 and Theorem V to get the result:

COROLLARY. The restriction that E be of finite measure in Theorem V may be omitted and the result will still be valid.

CHAPTER IV

ADDITIVE AND CLOSED TRANSFORMATIONS

§1

The purpose of this section is to introduce a number of notions.

DEFINITION 1. A transformation T from $\mathfrak{H}_1$ to $\mathfrak{H}_2$ is a single-valued function of the elements of $\mathfrak{H}_1$, which assumes values Tf in $\mathfrak{H}_2$. The set of f's for which Tf is defined is called the domain of T, the set of Tf's is called the range. The set $\mathfrak{T}$ of pairs $\{f,Tf\}$ in $\mathfrak{H}_1 \oplus \mathfrak{H}_2$ (Cf. Chapter III, §3, Theorem III) is called the graph of T. If T' is a transformation from $\mathfrak{H}_2$ to $\mathfrak{H}_1$, its graph is the set of pairs $\{T'g,g\}$.

LEMMA 1. A set S in $\mathfrak{H}_1 \oplus \mathfrak{H}_2$ is the graph of a transformation from $\mathfrak{H}_1$ to $\mathfrak{H}_2$ (from $\mathfrak{H}_2$ to $\mathfrak{H}_1$), if for a given $f \in \mathfrak{H}_1$, ($g \in \mathfrak{H}_2$) there is at most one pair of S having f (g) as its first (second) element.

T is obviously the transformation for which Tf is undefined if there is no pair of S with f as its first element and for which $Tf = g$ if $\{f,g\} \in S$.

DEFINITION 2. Let T be a transformation from $\mathfrak{H}_1$ to $\mathfrak{H}_2$ and let $\mathfrak{T}$ be the graph of T. Now if $\mathfrak{T}$ is also the graph of a transformation from $\mathfrak{H}_2$ to $\mathfrak{H}_1$, this second transformation T^{-1} is called the inverse of T.

The inverse of T when it exists, has for its domain and range, the range and domain of T. Also if f is in the domain of T, $T^{-1}(Tf) = f$ and for g in the range of T, $T(T^{-1}g) = g$.

DEFINITION 3. If T_1 and T_2 are two transformations from $\mathfrak{H}_1$ to $\mathfrak{H}_2$ such that $\mathfrak{T}_1 \subset \mathfrak{T}_2$, then T_1 is called a contraction of T_2 and T_2 an extension of T_1. We write this symbolically $T_1 \subset T_2$.

LEMMA 2. T_2 is an extension of T_1 if and only if for every f in the domain of T, T_2f is defined and $T_2f = T_1f$.

LEMMA 3. A transformation T is additive. (Cf. Chapter II, §3, Definition 2.) if and only if its graph $\mathfrak{T}$ is an additive set.

LEMMA 4. An additive set $\mathfrak{T} \subset \mathfrak{H}_1 \oplus \mathfrak{H}_2$ is the graph of a transformation from $\mathfrak{H}_1$ to $\mathfrak{H}_2$ if and only if $\{\theta_1, h\} \in \mathfrak{T}$ implies $h = \theta_2$.

By Lemma 1, the condition is necessary. It is also sufficient. For suppose $\{f, g_1\}$ and $\{f, g_2\}$ are in $\mathfrak{T}$. Then since $\mathfrak{T}$ is additive, $\{f, g_1\} - \{f, g_2\} = \{\theta_1, g_1 - g_2\} \in \mathfrak{T}$. Our condition implies $g_1 = g_2$ and thus there is at most one pair $\{f, g\} \in \mathfrak{T}$ with f in the first place.

DEFINITION 4. Let T be a transformation with graph $\mathfrak{T}$. If $\mathfrak{A}(\mathfrak{T})$ is the graph of a transformation, T_a, this latter transformation is called the additive extension of T.

DEFINITION 5. A transformation T from $\mathfrak{H}_1$ to $\mathfrak{H}_2$ will be said to be closed, if its graph is a closed set in $\mathfrak{H}_1 \oplus \mathfrak{H}_2$. If $[\mathfrak{T}]$ is the graph of a transformation $[T]$, $[T]$ is called the closure of T.

We note that $\mathfrak{M}(\mathfrak{T})$ is the graph of $[T_a]$ when this latter transformation exists. In general, given T, T_a will not exist. (A necessary and sufficient condition that T_a exist can be obtained by applying Lemma 4 to $\mathfrak{A}(\mathfrak{T})$). However even if T_a exists, $[T_a]$ need not exist. We give an example of this.

Let $\phi_1, \phi_2, \ldots$ be an orthonormal set. We define $T\phi_1 = \phi_1$.

Then it is easily seen that T_a exists. For if $\Sigma_{\alpha=1}^{n} a_\alpha \phi_\alpha = \theta$, then $a_\alpha = 0$ and hence $\Sigma_{\alpha=1}^{n} a_\alpha = 0$. Thus if $\{\theta, f\}$ is in $\mathfrak{U}(\mathfrak{T})$, $f = \theta$. Furthermore $T_a(\Sigma_{\alpha=1}^{n} a_\alpha \phi_\alpha) = (\Sigma_{\alpha=1}^{n} a_\alpha)\phi_1$. But $[T_\alpha]$ does not exist since $\{\theta, \phi_1\}$ is in $[\mathfrak{U}(\mathfrak{T})]$. To see this we notice that $T_a(\Sigma_{\alpha=1}^{n}(1/n)\phi_\alpha) = \phi_1$ and thus $\{\Sigma_{\alpha=1}^{n}(1/n)\phi_\alpha, \phi_1\}$ is in $\mathfrak{U}(\mathfrak{T})$. Now $|\Sigma_{\alpha=1}^{n}(1/n)\phi_\alpha| = (1/n)^{1/2} \longrightarrow 0$ as $n \longrightarrow \infty$. Hence $\{\Sigma_{\alpha=1}^{n}(1/n)\phi_\alpha, \phi_1\} \longrightarrow \{\theta, \phi_1\}$. Thus $\{\theta, \phi_1\}$ is in $[\mathfrak{U}(\mathfrak{T})]$ and the latter is not the graph of a transformation.

However Theorem II of Chapter II, §3, tells us that if T_a is a continuous transformation $[T_a]$ exists and has domain, the closure of the domain of T_a. Thus if a continuous additive T has domain, a linear manifold, T is closed.

DEFINITION 6. If T_1 is a transformation from $\mathfrak{H}_1$ to $\mathfrak{H}_2$ and T_2 is a transformation from $\mathfrak{H}_2$ to $\mathfrak{H}_3$ then T_2T_1 is the transformation whose domain consist of those f's for which $T_2(T_1f)$ is defined and has the value $T_2(T_1f)$, i.e., $(T_2T_1)f = T_1(T_2f)$.

Since a continuous function of a continuous function is continuous, it follows that if T_1 and T_2 are continuous, T_2T_1 is also continuous. If in addition T_1 and T_2 are additive with bounds C_1 and C_2 (Cf. Chapter II, §3, Theorem I.), then T_2T_1 has bound C_2C_1.

DEFINITION 7. If T_1 and T_2 are two transformations from $\mathfrak{H}_1$ to $\mathfrak{H}_2$, T_1+T_2 is the transformation, the domain of which is the set of those elements for which T_1f and T_2f are defined and for which $(T_1+T_2)f = T_1f+T_2f$. aT_1 is the transformation, whose domain is that of T_1 and for which $(aT_1)f = a(T_1f)$.

The sum of two continuous transformations is again continuous and if in particular T_1 and T_2 are additive with bounds C_1 and C_2, then the bound of the sum is $\leq C_1+C_2$.

§2

THEOREM I. If T is a transformation from $\mathfrak{H}_1$ to $\mathfrak{H}_2$, with graph $\mathfrak{T}$ and domain $\mathfrak{D}$ then $\mathfrak{T}^\perp$ is the graph of a transformation from $\mathfrak{H}_2$ to $\mathfrak{H}_1$, if and only if $\mathfrak{M}(\mathfrak{D}) = \mathfrak{H}_1$.

Proof: $\mathfrak{T}^\perp$ is the graph of a transformation from $\mathfrak{H}_2$ to $\mathfrak{H}_1$, if and only if $\{h_1, \theta_2\} \in \mathfrak{T}^\perp$ implies $h = \theta_1$. But $\{h, \theta_2\} \in \mathfrak{T}^\perp$ is equivalent to

$$0 = (\{h_1, \theta_2\}, \{f, Tf\}) = (h, f)$$

for every f in the domain of $\mathfrak{T}$. $\{h, \theta_2\} \in \mathfrak{T}^\perp$ is equivalent to $h \in \mathfrak{D}^\perp$. Thus $\mathfrak{T}^\perp$ is the graph of a transformation if and only if $h \in \mathfrak{D}^\perp$ implies $h = \theta_1$.

But $h \in \mathfrak{D}^\perp$ implies $h = \theta_1$, if and only if $\mathfrak{D}^\perp = \{\theta_1\}$. Thus $\mathfrak{T}^\perp$ is the graph of a transformation if and only if $\mathfrak{D}^\perp = \{\theta_1\}$. But $\mathfrak{D}^\perp = \theta_1$ is equivalent to $(\mathfrak{D}^\perp)^\perp = \{\theta_1\}^\perp = \mathfrak{H}_1$ by Theorem VII of Chapter II, §5. But since $(\mathfrak{D}^\perp)^\perp = (\mathfrak{M}(\mathfrak{D})^\perp)^\perp = \mathfrak{M}(\mathfrak{D})$, we see that $\mathfrak{D}^\perp = \{\theta_1\}$ is equivalent to $\mathfrak{M}(\mathfrak{D}) = \mathfrak{H}_1$. From a preceding statement we see that $\mathfrak{T}^\perp$ is the graph of a transformation if and only if $\mathfrak{M}(\mathfrak{D}) = \mathfrak{H}$.

If in particular T is additive, $\mathfrak{D}$ is additive and $\mathfrak{U}(\mathfrak{D}) = \mathfrak{D}$. Thus $\mathfrak{M}(\mathfrak{D}) = [\mathfrak{U}(\mathfrak{D})] = [\mathfrak{D}]$ and the statement $\mathfrak{M}(\mathfrak{D}) = \mathfrak{H}_1$ is equivalent to $\mathfrak{D}$ is dense. We have then:

COROLLARY. If T in Theorem I is also additive, then $\mathfrak{T}^\perp$ is the graph of a transformation, if and only if $\mathfrak{D}$ is dense.

DEFINITION 1. If T is a transformation from $\mathfrak{H}_1$ to $\mathfrak{H}_2$ and if $\mathfrak{T}^\perp$ is the graph of a transformation from $\mathfrak{H}_2$ to $\mathfrak{H}_1$, we will denote the latter transformation by $T^\perp$ and $-T^\perp$ by T^*.†

THEOREM II. Let T be such that $T^\perp$ exists. Then (a) A pair $\{g_1, g_2\} \in \mathfrak{H}_1 \oplus \mathfrak{H}_2$ is such that $T^\perp g_2 = g_1$ if

† If T is a transformation between two Banach spaces $\mathfrak{E}_1$ and $\mathfrak{E}_2$, $T^\perp$ can be regarded as a transformation from $\mathfrak{E}_2^*$ to $\mathfrak{E}_1^*$ (the conjugate spaces).

and only if for every f in the domain of T,

$$(f,g_1)+(Tf,g_2) = 0;$$

(b) A pair $\{g_1,g_2\} \in \mathfrak{H}_1 \oplus \mathfrak{H}_2$ is such that $T^*g_2 = g_1$ if and only if for every f in the domain of T,

$$(f,g_1) = (Tf,g_2).$$

Since $T^* = -T^\perp$, (a) and (b) are equivalent. Inasmuch as

$$(\{f,Tf\},\{g_1,g_2\}) = (f,g_1)+(Tf,g_2),$$

the condition in (a) is equivalent to $\{g_1,g_2\} \in \mathfrak{T}^\perp$.

COROLLARY. Let T be such that $T^\perp$ exists. Then
(a) A transformation T' is $\subset T^\perp$ if and only if for every f in the domain of T and every g in the domain of T'

$$(f,T'g)+(Tf,g) = 0.$$

(b) A transformation T' is $\subset T^*$, if and only if for every f in the domain of T and every g in the domain of T',

$$(f,T'g) = (Tf,g).$$

THEOREM III. Let T be a transformation from $\mathfrak{H}_1$ to $\mathfrak{H}_2$ for which $T^\perp$ exists, i.e., $\mathfrak{M}(\mathfrak{D}) = \mathfrak{H}$. Then $[T_a]$ exists if and only if $T^\perp$ (or T^*) has domain dense.

By Definitions 4 and 5, we see that $[T_a]$ exists if and only if $\mathfrak{M}(\mathfrak{T})$ is the graph of a transformation. But $\mathfrak{M}(\mathfrak{T}) = (\mathfrak{T}^\perp)^\perp$. (Cf. Chapter II, §6). Furthermore since $\mathfrak{T}^\perp$ is a linear manifold, $T^\perp$ is additive. Thus the corollary to Theorem I of this section states that "Domain of $T^\perp$ dense" is equivalent to "$(\mathfrak{T}^\perp)^\perp$ is the graph of the transformation." Since $(\mathfrak{T}^\perp)^\perp = \mathfrak{M}(\mathfrak{T})$, the first sentence in this paragraph shows that this latter statement is equivalent to " $[T_a]$ exists."

COROLLARY 1. $[T_a]$ exists if and only if $(T^\perp)^\perp$ $(= (T^*)^*)$ exists. When they exist $[T_a] = (T^\perp)^\perp$ $(= (T^*)^*)$.

COROLLARY 2. If T is a closed additive transformation with domain dense, $(T^{\perp})^{\perp}$ $(= (T^*)^*)$ exists and equals T.

Thus a closed additive transformation with a dense domain is symmetrically related to its perpendicular and to its adjoint. We will abbreviate "closed additive with a dense domain" to "c.a.d.d."

We call a continuous additive transformation whose domain is the full space a "linear" transformation. As we remarked before Definition 6, in §1, a linear transformation is closed.

THEOREM IV. If T is a continuous additive transformation, whose domain is dense and with bound C, (Cf. Chapter II, §3, Theorem I), then $T^{\perp}$ (and T^*) is a linear transformation with the same bound as T.

PROOF: $[T]$ exists by Theorem II of Chapter II, §3. Since $[T]^{\perp} = T^{\perp}$, we may suppose that $T = [T]$ and has domain the full space. By Theorem III of this section, $T^{\perp}$ has domain dense. It is also c.a. since $\mathfrak{T}^{\perp}$ is linear manifold. Thus Theorem II of Chapter II, §3, implies that $T^{\perp}$ is linear if it is continuous.

By Theorem I of Chapter II, §3, we have for every f $|Tf| \leqq C|f|$ (we have assumed that $[T] = T$). Hence for every f and g, $|(Tf,g)| \leqq |Tf|\cdot|g| \leqq C\cdot|f|\cdot|g|$. If g is in the domain of $T^{\perp}$, we have by (a) of Theorem II of this section that

$$(T^{\perp}g,f)+(g,Tf) = 0.$$

Hence

$$|(T^{\perp}g,f)| = |(g,Tf)| \leqq C\cdot|f|\cdot|g|.$$

If we let $f = T^{\perp}g$, we get $|T^{\perp}g|^2 \leqq C\cdot|T^{\perp}g|\cdot|g|$ which implies $|T^{\perp}g| \leqq C\cdot|g|$. Theorem I of Chapter II implies that $T^{\perp}$ is continuous, with a bound $C^{\perp} \leqq C$. Since however $(T^{\perp})^{\perp} = T$, we also have $C \leqq C^{\perp}$ and thus the bounds must be equal.

THEOREM V. If T_1 and T_2 are additive transformations with dense domains, then if $(T_2T_1)^*$ exists (or $(T_1+T_2)^*$), we have that $T_1^*T_2^*$ is a contraction of $(T_2T_1)^*$, $(T_1^*+T_2^*$ is

a contraction of $(T_1+T_2)^*$). $(aT_1)^* = \bar{a}T^*$ if $a \neq 0$.
If T_1 and T_2 are linear, we have that $T_1^* T_2^* = (T_2T_1)^*$.
(Similarly $T_1^*+T_2^* = (T_1+T_2)^*$).

PROOF: If f is in the domain of T_2T_1 and g in the domain of $T_1^*T_2^*$, (b) of Theorem II of this section implies

$$(T_2T_1f,g) = (T_1f,T_2^*g) = (f,T_1^*T_2^*g).$$

Now (b) of the Corollary of Theorem II of this section implies $T_1^*T_2^* \subset (T_2T_1)^*$. In the case of T_1+T_2 the argument is similar.

To show that $(aT_1)^* = \bar{a}T_1^*$, we note that if $a \neq 0$ $(T_1f,g_1) = (f,g_2)$ is equivalent to $(aT_1f,g_1) = (f,\bar{a}g_2)$.

If T_1 and T_2 are linear T_1^* and T_2^* are also by Theorem IV above. Thus $T_1^*T_2^*$ is everywhere defined and has no proper extensions and we must have $T_1^*T_2^* = (T_2T_1)^*$. This argument also applies to the sum.

COROLLARY. If T_1 is c.a.d.d. and T_2 is linear, then $(T_2T_1)^* = T_1^*T_2^*$.

PROOF: We know that $(T_2T_1)^* \supset T_1^*T_2^*$. On the other hand let f be in the domain of $(T_2T_1)^*$ and let g be in the domain of T_1 and hence in that of T_2T_1. Then

$$(g,(T_2T_1)^*f) = (T_2T_1g,f) = (T_1g,T_2^*f).$$

Since this holds for every g in the domain of T_1, we have that $T_1^*(T_2^*f)$ exists and equals $(T_2T_1)^*f$. This implies that $T_1^*T_2^* \supset (T_2T_1)^*$.

LEMMA 1. Let T be c.a.d.d. Let $\mathfrak{N}^*$ be the set of f's for which $T^*f = 0$. Let $\mathfrak{R}$ denote the range of T. Then $\mathfrak{R}^\perp = \mathfrak{N}^*$.

Since T^* is c.a., $\mathfrak{N}^*$ is closed. Since

$$(\{\theta_1,g\},\{f,Tf\}) = (g,Tf),$$

we see that $\{\theta,g\}$ is in $\mathfrak{T}^\perp$ if and only if $g \in \mathfrak{R}^\perp$. Thus $\mathfrak{R}^\perp$ is the set of zeros of $T^\perp = -T^*$.

It is evident geometrically that if T is c.a.d.d. and T^{-1} and T^{*-1} exist, then $(T^{-1})^* = T^{*-1}$. Lemma 4 of §1 and the preceding Lemma shows that T^{*-1} exists if and only if $[\mathfrak{R}] = \mathfrak{H}$ and that T^{-1} exists if and only if $[\mathfrak{R}^*] = \mathfrak{H}$.

THEOREM VI. Let $\mathfrak{N}$ denote the zeros of T, $\mathfrak{N}^*$ denote the zeros of T^*, $\mathfrak{R}$ the range of T, $\mathfrak{R}^*$ the range of T^*. Then $\mathfrak{N}^* = \mathfrak{R}^\perp$, $\mathfrak{N} = (\mathfrak{R}^*)^\perp$. T^{-1} exists if and only if $\mathfrak{N}^* = (\mathfrak{R}^*)^\perp = \{\theta\}$. T^{*-1} exists if and only if $\mathfrak{N}^* = (\mathfrak{R})^\perp = \{\theta\}$. If T^{-1} and T^{*-1} both exist, $(T^{-1})^* = T^{*-1}$.

§3

We now introduce certain notions which are fundamental in our discussion.

DEFINITION 1. An additive transformation H within $\mathfrak{H}$, will be called symmetric if (a) the domain of H is dense and (b) for every f and g in the domain of H,

$$(Hf,g) = (f,Hg).$$

From §2, Theorem I, we see that H^* exists. By (b) of the corollary to Theorem II of §2, we see that $H \subset H^*$. Thus we obtain the following Lemma.

LEMMA 1. An additive transformaion H is symmetric, if (a) it has domain dense and,(b) $H \subset H^*$.

LEMMA 2. If H is symmetric, $[H]$ exists and is symmetric.

PROOF: H^* is a closed transformation. Since $H \subset H^*$, we must have the graph of H in a closed set which is the graph of a transformation. Thus Lemma 1 of §1 of this Chapter, shows that the closure of the graph of H must be the graph of a transformation. Thus $[H]$ exists. From the graphs, it follows that $[H]^\perp = H^\perp$ and hence $[H]^* = H^*$.

Lemma 2 permits us in general to consider only closed symmetric transformations.

DEFINITION 2. If $H^* = H$, H is called self-adjoint.

LEMMA 3. A self-adjoint transformation is symmetric. If H is closed symmetric and H^* is symmetric, then H is self-adjoint. If the domain of a symmetric transformation H is the full space, H is self-adjoint. A symmetric linear transformation is self-adjoint.

The first sentence is a consequence of Lemma 1. If H is closed symmetric and H^* is symmetric, we obtain by Lemma 1 and Corollary 2 of Theorem III of the preceding section that $H \subset H^* \subset (H^*)^* = H$. The third statement follows from Lemma 1 of this section since a transformation with domain the full space can have no proper extension. The fourth statement follows from the third.

LEMMA 4. If H_1 and H_2 are symmetric and the domain of H_1+H_2 is dense, the latter transformation is symmetric. If a is real, aH_1 is symmetric and if H_1 is self-adjoint, a real, then aH_1 is self-adjoint.

This is a consequence of Theorem V of the preceding section. For if the domain of H_1+H_2 is dense, $(H_1+H_2)^*$ exists. Then too, $H_1+H_2 \subset H_1^*+H_2^* \subset (H_1+H_2)^*$ by this theorem. The second sentence is an immediate consequence.

LEMMA 5. If H_1 is self-adjoint and H_2 linear symmetric (and hence self-adjoint by Lemma 3 above) then H_1+H_2 is self-adjoint.

PROOF. The domain of H_1+H_2 is the same as that of H_1 and thus is dense. Hence Lemma 4, tells us that H_1+H_2 is symmetric and that $-H_2$ is self-adjoint. Furthermore $(H_1+H_2)+(-H_2)$ has domain the domain of H_1. Hence $H_1 = (H_1+H_2)+(-H_2) \subset (H_1+H_2)^*+(-H_2)^* \subset ((H_1+H_2)+(-H_2))^* = H_1^* = H_1$. This implies that the domain of $(H_1+H_2)^*$ which is the same as that of $(H_1+H_2)^*+(-H_2)^*$ is included in the domain of H_1 which is also the domain of H_1+H_2. Since H_1+H_2 is symmetric, $H_1+H_2 \subset (H_1+H_2)^*$. Since the domain of $(H_1+H_2)^*$ is included in that of H_1+H_2, we must have $(H_1+H_2)^* = H_1+H_2$.

LEMMA 6. If H is symmetric, and if H^{-1} exists, H^{-1} is symmetric if H^{*-1} exists, i.e., if $[\mathfrak{R}] = \mathfrak{H}$.

This is a consequence of Theorem VI of the preceding section. We note that if H^{*-1} exists, since $H \subset H^*$, we must have $H^{-1} \subset H^{*-1}$.

LEMMA 7. If H is self-adjoint and H^{-1} exists, then H^{-1} is self-adjoint.

This is a consequence of the last sentence of Theorem VI of the preceding section.

Another consequence of Theorem VI of the preceding section and $H \subset H^*$ is Lemma 8.

LEMMA 8. If H is closed symmetric, then $\mathfrak{N} \subset \mathfrak{N}^* = \mathfrak{R}^{\perp}$.

Suppose H_1 and H_2 are closed symmetric and $H_1 \subset H_2$. Then we have $H_1 \subset H_2 \subset H_2^* \subset H_1^*$. Now if H_1 is self-adjoint since $H_1 = H_1^*$, we see that H_2 must equal H_1 and H_1 has no proper symmetric extension. On the other hand, it is also conceivable that H_1 is symmetric with graph $\mathfrak{W}_1$ and $\mathfrak{W}_1^* \mathfrak{W}_1^{\perp}$ is one dimensional. If H_2 is then a symmetric closed extension of H_1, we have $H_1 \subset H_2 \subset H_1^*$. This last inclusion and Chapter II §5, Corollary 1 to Theorem VI imply that either $H_2 = H_1$ or $H_2 = H_1^*$. But H_1^* is not symmetric because $(H_1)^{**} = H_1$ is $\subset H_1^*$, but $H_1 \neq H_1^*$. Under these circumstances then H_1 would have no proper symmetric extension and yet not be self-adjoint. We shall show later the existence of an H_1 having these properties and give a complete discussion of this phenomena. But for the present, we simply introduce the definitions.

DEFINITION 3. If H_1 and H_2 are closed symmetric transformations such that $H_1 \subset H_2$, then H_2 is called a symmetric extension of H_1. If in addition $H_1 \neq H_2$, H_2 is called a proper symmetric extension of H_1. If H_1 is closed symmetric and has no proper symmetric extensions, H_1 is called maximal symmetric.

LEMMA 9. A self-adjoint transformation is maximal symmetric.

If H is symmetric and f is in the domain of H, then $(Hf,f) = (\overline{f,Hf}) = (\overline{Hf,f})$. Thus (Hf,f) is real and we may make the following definitions.

DEFINITION 4. Suppose H is symmetric. If there is a real number C such that for every f $(\neq \theta)$ in the domain of H, $C(f,f) \leqq (Hf,f)$, we let C_- be the least upper bound of such C's. Obviously C_- is such a C. If no C exists, let $C_- = -\infty$. If there is a real number C such that $(Hf,f) \leqq C(f,f)$ for every f $(\neq \theta)$ in the domain of H, we let C_+ be the greatest lower bound of such C's. Otherwise we write $C_+ = \infty$.

If $C_- \geqq 0$, we say that H is definite.

LEMMA 10. If $C = \max(|C_+|, |C_-|)$, is $< \infty$ then H is bounded with bound C.

We notice that for every f in the domain of H,

$$|(Hf,f)| \leqq C\cdot|f|^2.$$

If f and g are in the domain of H, then

$$(H(f\pm g),f\pm g) = (Hf,f)+(Hg,g)\pm 2R((Hf,g))$$

since $(Hg,f) = (g,Hf) = (\overline{Hf,g})$. Hence

$$R((Hf,g)) = \tfrac{1}{4}((H(f+g),f+g)-(H(f-g),f-g)),$$

This equation and the preceding inequality on C yields

$$|R(Hf,g)| \leqq \tfrac{1}{4}C(|f+g|^2+|f-g|^2)$$
$$= \tfrac{1}{2}C(|f|^2+|g|^2)$$

using B(11) of Chapter II, §1.

Now if $(Hf,g) = \zeta\cdot|(Hf,g)|$ where $|\zeta| = 1$, then $f' = \zeta^{-1}f$ is in the domain of H. Furthermore $|f'|^2 = |f|^2$ and $R((Hf',g)) = R(\zeta^{-1}(Hf,g)) = |(Hf,g)|$. Thus the last inequality of the preceding paragraph becomes

$$|(Hf,g)| \leqq \tfrac{1}{2}C(|f|^2+|g|^2).$$

This holds for every f and g in the domain of H. But since the domain of H is dense, we see by continuity that this inequality holds for every g.

Furthermore, if λ is real and not zero, we may let $f' = (1/\lambda)f$, $g' = \lambda g$ and the inequality becomes

$$|(Hf,g)| \leqq \tfrac{1}{2}C((1/\lambda^2)\cdot|f|^2+\lambda^2\cdot|g|^2)$$

for every real non-zero λ. If $f \neq \theta$ and $g \neq \theta$, we may let $\lambda^2 = |f|/|g|$ and obtain

$$|(Hf,g)| \leqq C\cdot|f|\cdot|g|.$$

If either $f = \theta$ or $g = \theta$, this last inequality is obvious. This inequality holds for every f in the domain of H and for every g. If we let $g = Hf$, we get

$$|Hf|^2 \leqq C\cdot|f|\cdot|Hf|$$

which implies $|Hf| \leqq C\cdot|f|$. Theorem I of Chapter II, §3, now gives the result.

§4

THEOREM VII. If T is c.a.d.d., then $(1+T^*T)^{-1}$ exists, is self-adjoint, has domain the full space and is definite and bounded with a bound $\leqq 1$.

PROOF. If $\{h,k\}$ is any pair of $\mathfrak{H}_1 \oplus \mathfrak{H}_2$, it can be expressed as the sum of an element of $\mathfrak{T}$ and an element of $\mathfrak{T}^{\perp}$ by Theorem VI of Chapter II, §5. Thus given h and k there is a unique f in the domain of T and a g in the domain of $T^{\perp}$ such that

$$\{h,k\} = \{f,Tf\}+\{T^{\perp}g,g\} = \{f,Tf\}+\{-T^*g,g\}$$

or such that

$$h = f-T^*g$$
$$k = Tf+g.$$

In particular if $k = \theta$, this means that to every H, there is an f in the domain of T such that

$$h = (1+T^*T)f\ .$$

This f is unique, since if there were two distinct f 's we would have two resolutions of $\{h,\theta\}$.

Thus for every h, $(1+T^*T)^{-1}h$ exists. $(1+T^*T)^{-1}$ is symmetric. The domain is dense and (b) of Definition 1 of the preceding section can be shown as follows. Let h and k be any two elements in $\mathfrak{H}$. Let $f = (1+T^*T)^{-1}h$, $g = (1+T^*T)^{-1}k$. f and g are in the domain of T^*T. Hence

$$(h,(1+T^*T)^{-1}k) = ((1+T^*T)f,g)$$
$$=(f,g)+(T^*Tf,g) = (f,g)+(Tf,Tg)$$
$$= (f,(1+T^*T)g) = ((1+T^*T)^{-1}h,k).$$

Lemma 3 of the preceding section now shows that $(1+T^*T)^{-1}$ is self-adjoint.

It is also definite and bounded with a bound $\leqq 1$. For

$$((1+T^*T)^{-1}h,h) = (f,(1+T^*T)f)$$
$$= (f,f)+(f,T^*Tf) = (f,f)+(Tf,Tf) = |f|^2+|Tf|^2 \geqq 0.$$

Thus $A = (1+T^*T)^{-1}$ is definite and furthermore

$$(Ah,h) \geqq (f,f) = (Ah,Ah) = |Ah|^2.$$

Now for every h, we must have

$$|Ah|\cdot|h| \geqq |(Ah,h)| \geqq |Ah|^2.$$

This implies $|h| \geqq |Ah|$.

THEOREM VIII. If T is c.a.d.d, T^*T is self-adjoint. If T' denotes contraction of T, with domain the domain of T^*T, then $[T'] = T$.

PROOF. By Theorem VII, $(1+T^*T)^{-1}$ is self-adjoint. By Lemma 7 of the preceding section, $1+T^*T$ is self-adjoint. If in Lemma 5, we let $H_1 = 1+T^*T$, $H_2 = -1$, we obtain that T^*T is self-adjoint.

It remains to prove our statement concerning T'. Since $T' \subset T$, we must have $[T'] \subset T$. If then $[T'] \neq T$, there must be a non-zero pair $\{g,Tg\}$ of $\mathfrak{T}$ which is orthogonal to all $\{f,Tf\}$ for which T^*Tf can be defined. (Cf. Corollary 1, to Theorem VI of Chapter II, §5). Thus for every f in the domain of T^*T,

$$0 = (\{g,Tg\},\{f,Tf\}) = (g,f)+(Tg,Tf)$$

$$= (g,f)+(g,T^*Tf) = (g,(1+T^*T)f).$$

But for every h in $\mathfrak{H}$, we can find an f in the domain of T^*T such that $h = (1+T^*T)f$, by Theorem VII of this section. Thus for every h in $\mathfrak{H}$, we have $(g,h) = 0$, and thus $g = \theta_1$, $Tg = \theta_2$, contrary to our assumption that $\{g,Tg\}$ is a non-zero pair. This contradiction shows that $[T'] = T$.

COROLLARY. Theorems VII and VIII hold if T^* is written in place of T, T in place of T^*.

This is a consequence of Corollary 2 of Theorem III of §2 of this Chapter, since this result permits us to substitute T^* for T.

CHAPTER V

WEAK CONVERGENCE

§1

In this section, we shall discuss the weak convergence of elements in Hilbert space. This notion applies in more general spaces as we shall indicate.

DEFINITION. A sequence of elements $\{f_n\}$ of $\mathfrak{H}$, will be said to be weakly convergent, if to every $g \in \mathfrak{H}$, the $\lim_{n\to\infty}(f_n,g)$ exists.*

We shall establish for every weakly convergent sequence $\{f_n\}$ the existence of an $f \in \mathfrak{H}$, such that for every g, $(f_n,g) \longrightarrow (f,g)$.

LEMMA 1. Let T_n be any sequence of continuous additive functions, whose domain is the full space $\mathfrak{H}$ and whose values are in a linear space. Then if there is a sphere $\mathfrak{K}$ and a constant C such that for $f \in \mathfrak{K}$, $|T_nf| \leqq C(>0)$, then the T_n 's are uniformly bounded.

We know from Theorem I, of Chapter II, §3, that to every T_n we have a C_n such that $|T_nf| \leqq C_n|f|$ for every f. We must show that the C_n 's are bounded. Suppose that they are not. Then if r is the radius of $\mathfrak{K}$, it must be possible to find a C_n such that $C_n > 6C/r$. Then given ϵ, it is possible to find an f $(= f_\epsilon)$ such that $|T_nf| > (1-\epsilon)C_n\cdot|f|$ and we may take $|f| = r/2$. Thus if f_0 is the center of $\mathfrak{K}$, f_0+f is in $\mathfrak{K}$ and we must have $|T_n(f_0+f)| \leqq C$. Hence $|T_nf_0+T_nf| < C$, which implies $|T_nf| \leqq 2C$ since $|T_nf_0| < C$. But $|T_nf| \geqq (1-\epsilon)C_n|f| = \frac{1}{2}(1-\epsilon)C_n\cdot r \geqq (1-\epsilon)3C$. Now $(1-\epsilon)3C$ cannot remain less than $2C$ for every $\epsilon > 0$ and thus we have a con-

---

* In general Banach space, a sequence of elements is said to be weakly convergent if for every linear functional, $F, F(f_n)$ is convergent.

tradiction. Hence the C_n 's are uniformly bounded.

THEOREM I. Let T_n be a sequence of continuous additive functions, whose domain is the full space $\mathfrak{H}$ and whose values lie in a linear space. Suppose that for every f in $\mathfrak{H}$, $T_n f$ is convergent. Then the bounds of T_n 's are bounded.

PROOF: Let us suppose that the Theorem does not hold for a specific sequence $\{T_n\}$. Then Lemma 1 above implies that the $|T_n f|$ are unbounded in every sphere.

Now suppose that for $i = 1, \ldots, k$ we have specified a function T_{n_i}, a sphere $\mathfrak{K}_i$, with a center f_i and radius r_i and such that if $f \in \mathfrak{K}_i$, $|T_{n_i} f| \geqq i$. Suppose also that $r_i \leqq \frac{1}{2} r_{i-1} \leqq 1/2^i$ and $\mathfrak{K}_{i+1} \subset \mathfrak{K}_i$.

We know that the $T_n f$'s are not bounded in the sphere with center f_k and radius $\frac{1}{2} r_k$. We can therefore find a $T_{n_{k+1}}$ and an f_{k+1} within this sphere, with $|T_{n_{k+1}} f_{k+1}| \geqq 2(k+1)$. Since $T_{n_{k+1}}$ is continuous, we can find a closed sphere $\mathfrak{K}_{k+1}$ with center f_{k+1} and radius $r_{k+1} \leqq \frac{1}{2} r_k \leqq 1/2^{k+1}$, for which $|T_{n_{k+1}} f| \geqq k+1$ for $f \in \mathfrak{K}_{k+1}$. If g is in $\mathfrak{K}_{k+1}$,

$$|f_k - g| \leqq |f_k - f_{k+1} + f_{k+1} - g| \leqq |f_k - f_{k+1}| + |f_{k+1} - g| \leqq r_k$$

or g in $\mathfrak{K}_k$. Hence $\mathfrak{K}_{k+1} \subset \mathfrak{K}_k$ and we see that we may define a sequence of T_{n_i}, $\mathfrak{K}_i$, f_i, r_i, which have the properties given in the preceding paragraph for every i.

Since each $\mathfrak{K}_i$ contains all that follow and $r_n \longrightarrow 0$ as $n \longrightarrow \infty$, the f_n 's form a convergent sequence, whose limit f is in every sphere $\mathfrak{K}_i$. Consequently $|T_{n_i} f| \longrightarrow \infty$ as $i \longrightarrow \infty$ and the $T_{n_i} f$'s cannot converge. This contradiction shows that the T_n 's must be uniformly bounded.

COROLLARY 1. If in Theorem I, for each f, $|T_n f|$ is bounded, then the result still holds.

COROLLARY 2. A weakly convergent sequence of elements $\{f_n\}$ must have the norms $\{\|f_n\|\}$ bounded.

Let f_n be a weakly convergent sequence. We have a C such that $|f_n| \leqq C$. Now for every g, $(g,f_n) \longrightarrow F(g)$, where $F(g)$ denotes the value of the limit. Since $|(g,f_n)| \leqq |f_n| \cdot |g| \leqq C \cdot |g|$ for every g, we have $|F(g)| \leqq C \cdot |g|$. Since F is obviously additive, Theorem I of Chapter II, §3, implies that F is a linear functional. Thus there is a $f \in \mathfrak{H}$ such that $(g,f) = F(g)$ for every g, by Theorem IV of Chapter II, §4. Thus we have established:

THEOREM II. If $\{f_n\}$ is a weakly convergent sequence of elements of $\mathfrak{H}$, there exists an f such that for every g in $\mathfrak{H}$, $(g,f_n) \longrightarrow (g,f)$.

§2

Thus if a sequence $\{f_n\}$ is weakly convergent it has a weak limit f, i.e., $(f_n,g) \longrightarrow (f,g)$ for every g. Thus $\mathfrak{H}$ is complete for weak convergence too.

Since (f,g) is continuous in f, we have

LEMMA 1. If a sequence $\{f_n\}$ is strongly convergent to f, it is weakly convergent to the same limit.

The converse of this lemma does not hold. For let $\phi_1, \phi_2, \ldots$ be an infinite orthonormal set. For every g we have $\Sigma_{\alpha=1}^{\infty} |a_\alpha|^2 < \infty$, where $a_\alpha = (g,\phi_\alpha)$. Thus for every g, $(g,\phi_\alpha) \longrightarrow 0$ and the ϕ_α's form a weakly convergent series. Since however $|\phi_\alpha - \phi_\beta| = \sqrt{2}$ if $\alpha \neq \beta$, they are not strongly convergent.

This example also shows that there are bounded infinite sets of elements, which have no limit points. Thus Hilbert space is not locally compact. However for weak convergence, we have a kind of compactness.

THEOREM III. If $\{f_n\}$ is a bounded sequence of elements, there exists a weakly convergent subsequence.

PROOF: Let $g_1, g_2, \ldots$ be a denumerable set, dense in $\mathfrak{H}$. The numbers (g_1,f_α) are bounded and thus we can find a subsequence $\{f'_\alpha\}$ for which (g_1,f'_α) is convergent. Similarly, we can chose a subsequence $\{f''_\alpha\}$ of $\{f'_\alpha\}$ such that (g_2,f''_α) is convergent. By this process, we can continue to choose subsequences so that $(g_1, f_\alpha^{(n)})$ is convergent for $i \leqq n$. The "diagonal sequence" $\{f_\alpha^{(\alpha)}\}$ then has the property that for each n (when the first n elements are ignored) it is a subsequence of $\{f_\alpha^{(n)}\}$. Hence $(g_1, f_\alpha^{(\alpha)})$ is convergent for every i.

The norms of the $\{f_\alpha^{(\alpha)}\}$ are bounded and thus the linear functionals $(g, f_\alpha^{(\alpha)})$ are uniformly continuous on every bounded region. Since these functionals also converge on a dense set, they must converge for every value of g.

§3

Our purpose in this section is to prove Theorems IV and V below. For this, we prove the following lemmas.

LEMMA 1. Let T be a linear transformation from $\mathfrak{H}_1$ to $\mathfrak{H}_2$. The domain of T is the full space and we let $\mathfrak{K}_n$ denote the set in $\mathfrak{H}_2$ of those elements in the form Tf, $|f| \leqq n$, $|Tf| \leqq 1$. The set $\mathfrak{K}_n$ is closed.

Since T is linear, T^* is also linear. (Cf. Chapter IV, §2, Theorem IV) Now let g be a limit point of $\mathfrak{K}_n$. We can find a sequence $\{g_1\}$ with $g_1 \longrightarrow g$ and such that $g_1 = Tf_1$ for an f_1 with $|f_1| \leqq n$. Since the f's are uniformly bounded we can find a subsequence $\{f_n'\}$, which converges weakly to an f with $|f| \leqq n$ (Cf. Theorem III, §2 above). Let $g_1' = Tf_1'$. Then for every h,

$$
\begin{aligned}
(f, T^*h) &= \lim_{\alpha \to \infty} (f_\alpha', T^*h) \\
&= \lim_{\alpha \to \infty} (Tf_\alpha', h) \\
&= \lim_{\alpha \to \infty} (g_\alpha', h) \\
&= (g, h).
\end{aligned}
$$

Since $(T^*)^* = T$, (b) of Theorem II of Chapter IV, §2 implies $Tf = g$. Since $|f| \leqq n$, $|g| \leqq 1$, g is in $\mathfrak{K}_n$. Thus $\mathfrak{K}_n$ is closed.

LEMMA 2. Let T be a linear transformation fromm $\mathfrak{H}_1$ to $\mathfrak{H}_2$ for which T^{-1} exists. Let $\mathfrak{K}_n$ be as in Lemma 1. Then if for some $n = n_0$, $\mathfrak{K}_n$ contains a sphere $\mathfrak{K}$, then

T^{-1} is bounded.

Let $\mathfrak{K}$ have radius r and center g_1. For $g \in \mathfrak{K} \subset \mathfrak{K}_n$, we have that $g = Tf$ for an f with $|f| \leqq n$. In particular this is true for $g_1 = Tf_1$. Now if h is such that $|h| < r$, we have that $g = g_1+h$ is in $\mathfrak{K}$ and thus $h = g-g_1 = T(f-f_1)$. Since $|f-f_1| \leqq |f|+|f_1| \leqq 2n$, h is in $\mathfrak{K}_{2n}$. Thus $\mathfrak{K}_{2n}$ contains the sphere with center θ_2 and radius r.

This implies that T^{-1} is defined everywhere and has a bound $\leqq 4n/r$. For if $g \in \mathfrak{H}_2$ and $g \neq \theta_2$, let $h = (r/2|g|)g$. Since $h \in \mathfrak{K}_{2n}$, (Cf. above) we have that $|T^{-1}h| \leqq 2n$, which is equivalent to $|T^{-1}g| \leqq (4n/r)\cdot|g|$.

We introduce certain set-theoretic definitions which have played a very important rôle in the general theory of linear spaces.

DEFINITION 1. A set S is said to be nowhere dense if every sphere contains a sphere of the complement of S. A set will be said to be of the first category if it is a denumerable sum of nowhere dense sets.

The following Lemma is important.

LEMMA 3. A set S of the first category does not contain any sphere.

Let $S = S_1+S_2+ \ldots$ where S_i is nowhere dense. Let $\mathfrak{K}$ be any sphere. We shall show that S does not contain $\mathfrak{K}$. Within $\mathfrak{K}$, we can find a $\mathfrak{K}_1$ belonging to the complement of S_1 and we can suppose that the radius r_1 of $\mathfrak{K}_1$ is $\leqq 1/2$. Within $\mathfrak{K}_1$, we can find a sphere $\mathfrak{K}_2$ of the complement of S_2 with radius $r_2 \leqq 1/2^2$. Continuing, we can find a sequence of spheres $\{\mathfrak{K}_i\}$ each containing all subsequent spheres, $\mathfrak{K}_i$ in the complement of S_i and whose radii approach zero. The centers $\{f_i\}$ of the spheres $\{\mathfrak{K}_i\}$ form a convergent series, with a limit f, which is in every sphere including $\mathfrak{K}$. Since f is in every sphere, $\mathfrak{K}_i$, it is in the complement of every S_i and hence in the complement of S. Since f is in $\mathfrak{K}$, , S does not contain $\mathfrak{K}$.

LEMMA 4. Let T be a linear transformation from $\mathfrak{H}_1$ to $\mathfrak{H}_2$ for which T^{-1} exists. Let $\mathfrak{K}_n$ be as in Lemmas 1 and 2. If T^{-1} is not bounded, each $\mathfrak{K}_n$ is nowhere dense.

PROOF: Let $\mathfrak{K}$ be any sphere of $\mathfrak{H}_2$. $\mathfrak{K}_n$ does not contain $\mathfrak{K}$ by Lemma 2. Thus $\mathfrak{K}$ contains a point f of the complement of $\mathfrak{K}_n$. Since the complement of $\mathfrak{K}_n$ is open by Lemma 1 and its intersection with $\mathfrak{K}$ is not empty, $\mathfrak{K}$ and the complement of $\mathfrak{K}_n$ must have a sphere in common. Thus $\mathfrak{K}_n$ is nowhere dense.

THEOREM IV. Let T be a linear transformation from $\mathfrak{H}_1$ to $\mathfrak{H}_2$ whose inverse T^{-1} exists. Then if T has $\mathfrak{H}_2$ as its range T^{-1} is bounded.

Under these circumstances, the sum of the $\mathfrak{K}_n$ of Lemmas 1, 2, and 4 contains the unit sphere of $\mathfrak{H}_2$. If T^{-1} were not bounded then this sum would be of the first category by Lemma 4 and hence could not contain a sphere by Lemma 3. Thus T^{-1} is bounded.

THEOREM V. If T is a closed transformation whose domain is a closed linear manifold, then T is bounded.

PROOF: We consider A the transformation from $\mathfrak{T}$ $(=\mathfrak{H}_1)$ to $\mathfrak{D}$ $(= \mathfrak{H}_2)$, defined by the equation $A\{f,Tf\} = f$. A has domain $\mathfrak{T}$ and bound 1. Thus A is linear. The inverse A^{-1} exists and we note $A^{-1}f = \{f,Tf\}$. The range of A is $\mathfrak{H}_2$. Thus we may apply Theorem IV and obtain that there is a C such that for every $f \in \mathfrak{D}$,

$$C\cdot|f| \geqq |\{f,Tf\}| = (|f|^2+|Tf|^2)^{1/2} \geqq |Tf|.$$

CHAPTER VI

PROJECTIONS AND ISOMETRY

In this chapter we will consider four special kinds of transformations of particular interest in the theory that follows.

§1

DEFINITION 1. Let $\mathfrak{M}$ be a linear manifold of $\mathfrak{H}$ and for every f, let $f = f_1+f_2$, $f_1 \in \mathfrak{M}_1$, $f_2 \in \mathfrak{M}^{\perp}$. (Cf. Chapter II, §5, Theorem VI) The transformation E which is defined by the equation $Ef = f_1$ is called a projection.

Lemma 1. E is a linear self-adjoint transformation with $C_- \geqq 0$, $C_+ \leqq 1$. (Cf. Def. 4 of Chapter IV, §3) Furthermore $E^2 = E$.

PROOF: The uniqueness of the resolution of Theorem VI of Chapter II, is easily seen to imply that $E^2 = E$ and that E is additive. We also see from the same Theorem that the domain of E is $\mathfrak{H}$. Since $|f|^2 = |f_1|^2+|f_2|^2$, $|f| \geqq |Ef|$ and thus E is bounded. Hence E is linear.

Now for f and $g \in \mathfrak{H}$, we resolve $f = f_1+f_2$, $g = g_1+g_2$ and then since f_1 and g_1 are orthogonal to f_2 and g_2 we obtain,

$$(Ef,g) = (f_1,g_1+g_2) = (f_1,g_1) = (f_1+f_2,g_1) = (f,Eg).$$

Thus E is symmetric. Since it is also linear, we know by Lemma 3 of §3 of Chapter IV that E is self-adjoint. Since we also have

$$(Ef,f) = (f_1,f) = (f_1,f_1) = |f_1|^2 \leqq |f|^2$$

we see that $C_- \geqq 0$, $C_+ \leqq 1$. If both $\mathfrak{M}$ and $\mathfrak{M}^{\perp}$ are not $\{\theta\}$, we easily verify that $C_+ = 1$ and $C_- = 0$.

Conversely, we have

LEMMA 2. If $E^2 = E$ and E is closed symmetric then E is a projection.

Since $E^2 = E$, $\mathfrak{N}$ the set of zeros of E, includes all elements g for which $g = (1-E)f$, f in the domain of E. For since f is in the domain of E, and thus $Ef = E^2f = E(Ef)$, Ef must also be in the domain of E. Hence g is in the domain of E and $Eg = E(1-E)f = (E-E^2)f = \theta$.

By Chapter IV, §3, Lemma 8, we know that $\mathfrak{N} \subset \mathfrak{N}^* = [\mathfrak{R}]^{\perp}$. Thus $\mathfrak{N}$ and $\mathfrak{R}$ are orthogonal and for f in the domain of E,

$$f = Ef+(1-E)f$$
$$|f|^2 = |Ef|^2+|(1-E)f|^2.$$

Thus $|Ef| \leqq |f|$ and E is bounded. Since E is closed continuous and has domain dense, E is linear. (Cf. the definition of "linear" in Chapter IV, §2 preceding Theorem IV) Thus E has domain the full space.

Let $\mathfrak{M} = [\mathfrak{R}]$. For every f, E gives a resolution $f = f_1+f_2$, $Ef \in \mathfrak{M}$, $(1-E)f \in \mathfrak{M}^{\perp}$. Since only one such resolution is possible, it follows from the definition of projection, that E is the projection on $\mathfrak{M}$.

COROLLARY. If E is c.a.d.d. and $E^2 = E$ and $\mathfrak{R}$ is orthogonal to $\mathfrak{N}$ then E is a projection.

LEMMA 3. If E is a projection with range $\mathfrak{M}$, then $1-E$ is a projection with range $\mathfrak{M}^{\perp}$.

PROOF: $1-E$ is self-adjoint and $(1-E)(1-E) = 1-2E+E^2 = 1-E$. Lemma 2 implies that $1-E$ is a projection.

Now $Ef = \theta$ if and only if $f \in \mathfrak{M}^{\perp}$. Also $(1-E)f = f$ if and only if f is in the range of $1-E$, since $1-E$ is a projection. But $(1-E)f = f$ is equivalent to $Ef = \theta$ and thus $\mathfrak{M}^{\perp}$ is the range of $1-E$.

LEMMA 4. If E_1 is a projection with range $\mathfrak{M}_1$ and E_2 is a projection with range $\mathfrak{M}_2$, then E_1E_2 is a projection if and only if $E_1 \cdot E_2 = E_2 \cdot E_1$. If E_1E_2 is a projection, its range is $\mathfrak{M}_1 \cdot \mathfrak{M}_2$.

The condition $E_1E_2 = E_2E_1$ is by Theorem V of Chapter IV, §2, a necessary and sufficient condition that E_1E_2 be self-adjoint. Thus the condition of our Lemma is necessary and when it holds we know that E_1E_2 is self-adjoint. When it holds, we also have $(E_1E_2)^2 = E_1E_2E_1E_2 = E_1E_1E_2E_2 = E_1E_2$. Lemma 2 shows then that E_1E_2 is a projection.

If $E_1E_2 = E_2E_1$, then the range of E_1E_2 is in both that of E_1 and that of E_2, i.e., in $\mathfrak{M}_1 \cdot \mathfrak{M}_2$. Also if $f \in \mathfrak{M}_1 \cdot \mathfrak{M}_2$, $E_1E_2f = E_1(E_2f) = E_1f = f$, i.e., f is in the range of E_1E_2. Thus the range of E_1E_2 is $\mathfrak{M}_1 \cdot \mathfrak{M}_2$, when E_1E_2 is a projection.

LEMMA 5. If $E_1, \dots, E_n$ are n projections with ranges $\mathfrak{M}_1, \dots, \mathfrak{M}_n$ respectively, then $E_1 + \dots + E_n$ is a projection, if and only if $E_iE_j = 0$ if $i \neq j$. If $E_1 + \dots + E_n$ is a projection, then $\mathfrak{M}_i$ is orthogonal to $\mathfrak{M}_j$ for $i \neq j$ and the range of $E_1 + \dots + E_n$ is $\mathfrak{U}(\mathfrak{M}_1 \cup \dots \cup \mathfrak{M}_n)$ ($\cup$ indicating logical sum).

Let us suppose that $E_1 + \dots + E_n$ is a projection. Suppose $E_iE_j \neq 0$ for $i \neq j$. Hence there is an f such that $E_iE_jf \neq \theta$. Let $g = E_jf$. Then $g \in \mathfrak{M}_j$ and $E_ig \neq \theta$. Hence

$$|g|^2 \geqq ((E_1 + \dots + E_n)g, g) = \Sigma_{\alpha=1}^{n}(E_\alpha g, g) = \Sigma_{\alpha=1}^{n}|E_\alpha g|^2$$
$$\geqq |E_ig|^2 + |E_jg|^2 = |E_ig|^2 + |g|^2 > |g|^2.$$

This contradiction indicates that $E_iE_j = 0$.

On the other hand if $E_iE_j = 0$ for $i \neq j$, then $(E_1 + \dots + E_n)^2 = E_1 + \dots + E_n$. Since $E_1 + \dots + E_n$ is self-adjoint, it is a projection by Lemma 2 above.

If $E_iE_j = 0$, $E_j = E_j - E_iE_j = (1 - E_i)E_j$. It follows from Lemmas 3 and 4 above that $\mathfrak{M}_j$ is in $\mathfrak{M}_i^\perp \mathfrak{M}_j$ or $\mathfrak{M}_j$ in $\mathfrak{M}_i^\perp$. Thus $\mathfrak{M}_j$ is orthogonal to $\mathfrak{M}_i$.

Obviously the range of $E_1 + \dots + E_n$ must be in $\mathfrak{U}(\mathfrak{M}_1 \cup \dots \cup \mathfrak{M}_n)$. On the other hand if $f \in \mathfrak{U}(\mathfrak{M}_1 \cup \dots \cup \mathfrak{M}_n)$, it is readily established that $f = f_1 + \dots + f_n$, where $f_1 \in \mathfrak{M}_1, \dots, f_n \in \mathfrak{M}_n$. Now $(E_1 + \dots + E_n)(f_1 + \dots + f_n) = f_1 + \dots + f_n$ since $E_if_j = E_iE_jf_j = \theta$, if $i \neq j$. Thus $f_1 + \dots + f_n$ is in the range of $E_1 + \dots + E_n$ and $\mathfrak{U}(\mathfrak{M}_1 \cup \dots \cup \mathfrak{M}_n)$ is included in this range. This and the previous result prove the last statement of the lemma.

DEFINITION 2. Two projections E_1 and E_2 are called orthogonal if $E_iE_j = 0$, or what is equivalent, if $\mathfrak{M}_1$ is orthogonal to $\mathfrak{M}_2$.

LEMMA 6. If E_1 and E_2 are projections with ranges $\mathfrak{M}_1$ and $\mathfrak{M}_2$, then E_1-E_2 is a projection if and only if $E_2 = E_1E_2$. If E_1-E_2 is a projection then $\mathfrak{M}_2 \subset \mathfrak{M}_1$, and the range of E_1-E_2 is $\mathfrak{M}_1 \cdot \mathfrak{M}_2^{\perp}$.

It follows from Lemma 3, that E_1-E_2 is a projection if and only if $1-(E_1-E_2) = 1-E_1+E_2$ is a projection. By Lemma 5, $(1-E_1)+E_2$ is a projection if and only if $E_2(1-E_1) = 0$ or $E_2 = E_2E_1$. These two results imply the first statment of the Lemma.

If $E_2 = E_2E_1$, Lemma 4 implies $\mathfrak{M}_2 = \mathfrak{M}_1\mathfrak{M}_2$ or $\mathfrak{M}_2 \subset \mathfrak{M}_1$. Since $E_1-E_2 = E_1-E_2E_1 = (1-E_2)E_1$ Lemmas 3 and 4 imply that the range of E_1-E_2 is $\mathfrak{M}_2^{\perp}\mathfrak{M}_1$.

LEMMA 7. Let $E_1, E_2, \ldots$ be a sequence of mutually orthogonal projections, with range $\mathfrak{M}_1, \mathfrak{M}_2, \ldots$ respectively. Let $E = \Sigma_{\alpha=1}^{\infty} E_\alpha$ (i.e., $Ef = \lim_{n\to\infty} \Sigma_{\alpha=1}^{n} E_\alpha f$ whenever this limit exists). Then E is a projection with range $\mathfrak{M}(\mathfrak{M}_1 \cup \mathfrak{M}_2 \cup \ldots)$ (where $\cup$ denotes the logical sum).

PROOF: We note that if $n \geqq m$ and $f \in \mathfrak{H}$ then by Lemma 5

$$|f|^2 \geqq |(\Sigma_{\alpha=m+1}^{n} E_\alpha)f|^2 = |\Sigma_{\alpha=1}^{n} E_\alpha f - \Sigma_{\alpha=1}^{m} E_\alpha f|^2$$
$$= |\Sigma_{\alpha=m+1}^{n} E_\alpha f|^2 = \Sigma_{\alpha=m+1}^{n} |E_\alpha f|^2.$$

If we let $m = 0$, we obtain $|f|^2 \geqq \Sigma_{\alpha=1}^{n} |E_\alpha f|^2$. Hence $|f|^2 \geqq \Sigma_{\alpha=1}^{\infty} |E_\alpha f|^2$. This implies that $\Sigma_{\alpha=m+1}^{n} |E_\alpha f|^2 \longrightarrow 0$ as m and $n \longrightarrow \infty$. Our first inequality then shows that $|\Sigma_{\alpha=1}^{n} E_\alpha f - \Sigma_{\alpha=1}^{m} E_\alpha f| \longrightarrow 0$ as m and $n \longrightarrow \infty$. This shows that Ef exists for every f.

Now E is symmetric, since for every f and g of $\mathfrak{H}$,

$$(Ef,g) = (\lim_{n\to\infty} \Sigma_{\alpha=1}^{n} E_\alpha f, g) = \lim_{n\to\infty} ((\Sigma_{\alpha=1}^{n} E_\alpha)f, g)$$
$$= \lim_{n\to\infty} (f, (\Sigma_{\alpha=1}^{n} E_\alpha)g) = (f, (\lim_{n\to\infty} \Sigma_{\alpha=1}^{n} E_\alpha)g) = (f, Eg).$$

Furthermore

$$(\Sigma_{\alpha=1}^{\infty} E_\alpha)(\Sigma_{\beta=1}^{\infty} E_\beta) = \lim_{n\to\infty} (\Sigma_{\alpha=1}^{n} E_\alpha)(\Sigma_{\beta=1}^{\infty} E_\beta)$$
$$= \lim_{n\to\infty} (\Sigma_{\alpha=1}^{n} E_\alpha)(\lim_{m\to\infty} \Sigma_{\beta=1}^{m} E_\beta) = \lim_{n\to\infty} \lim_{m\to\infty} (\Sigma_{\alpha=1}^{n} E_\alpha)(\Sigma_{\beta=1}^{m} E_\beta)$$
$$= \lim_{n\to\infty} \lim_{m\to\infty} (\Sigma_{\alpha=1}^{n} E_\alpha) = \lim_{n\to\infty} \Sigma_{\alpha=1}^{n} E_\alpha = \Sigma_{\alpha=1}^{\infty} E_\alpha.$$

Hence Lemma 2 above shows that E is a projection.

We now prove the last statement of the Lemma. Let $\mathfrak{N}$ denote the set of elements f of $\mathfrak{H}$ in the form $f = \Sigma_{\alpha=1}^{\infty} f_\alpha$, where $f_\alpha \in \mathfrak{M}_\alpha$. Since the $\mathfrak{M}_\alpha$'s are orthogonal, $|f|^2 = \Sigma_{\alpha=1}^{\infty} |f_\alpha|^2$. Now a proof similar to the completeness argument of Theorem I of Chapter III, will show that $\mathfrak{N}$ is a closed set. Furthermore $\mathfrak{U}(\mathfrak{M}_1 \cup \mathfrak{M}_2 \cup \ldots) \subset \mathfrak{N}$ and every $f \in \mathfrak{N}$ is the limit of elements of $\mathfrak{U}(\mathfrak{M}_1 \cup \mathfrak{M}_2 \cup \ldots)$. Thus $\mathfrak{N}$ is the closure of $\mathfrak{U}(\mathfrak{M}_1 \cup \mathfrak{M}_2 \cup \ldots)$ that is $\mathfrak{M}(\mathfrak{M}_1 \cup \mathfrak{M}_2 \cup \ldots)$. One can easily verify that $\mathfrak{N}$ is the range of E and this completes the proof of the Lemma.

DEFINITION 3. We will write $E_2 \leqq E_1$, if $\mathfrak{M}_2 \subset \mathfrak{M}_1$.

LEMMA 8. The following statements are equivalent:

(a) $E_2 \leqq E_1$;

(b) $E_2E_1 = E_2$ $(= E_1E_2$ by Lemma 4);

(c) For every f of $\mathfrak{H}$, $|E_2f|^2 \leqq |E_1f|^2$.

PROOF: (a) implies (b). For if $f \in \mathfrak{H}$, we know that $f = f_1+f_2$ where $f_1 \in \mathfrak{M}_1$, $f_2 \in \mathfrak{M}_1^\perp$ By Corollary 1 to Theorem VI of Chapter II, §5, we have $f_1 = f_{1,1}+f_{1,2}$, where $f_{1,1} \in \mathfrak{M}_2$, $f_{1,2} \in \mathfrak{M}_2^\perp \cdot \mathfrak{M}_1$. Thus $f = f_{1,1}+(f_{1,2}+f_2)$. Since $\mathfrak{M}_1^\perp \subset \mathfrak{M}_2^\perp$ by Corollary 2 to Theorem VI of Chapter II, we have $f_2 \in \mathfrak{M}_2^\perp$. Thus $f_{1,2}+f_2 \in \mathfrak{M}_2^\perp$, $f_{1,1} \in \mathfrak{M}_2$ and these imply $E_2f = f_{1,1}$. But by the method of definition of $f_{1,1}$, $E_2E_1f = f_{1,1}$. Hence $E_2E_1f = E_2f$ and (a) implies (b).

(b) implies (c) since $|E_2f|^2 = |E_2E_1f|^2 \leqq |E_1f|^2$.

(c) implies (a). For if $f \in \mathfrak{M}_2$, $|f|^2 = |E_2f|^2 \leqq |E_1f|^2 \leqq |f|^2$. Thus $|f|^2 = |E_1f|^2$. Since $|(1-E_1)f|^2 = |f|^2-|E_1f|^2 = 0$, we must have $(1-E_1)f = \theta$ or $f = E_1f \in \mathfrak{M}_1$. Thus $\mathfrak{M}_2 \subset \mathfrak{M}_1$.

LEMMA 9. If E_1, E_2, ... denotes a sequence of projections with ranges $\mathfrak{M}_1, \mathfrak{M}_2$, ... respectively and such that $E_\alpha \leqq E_{\alpha+1}$, then $E = \lim_{n\to\infty} E_n$ is a projection such that $E_\alpha \leqq E$. The range of E is $\mathfrak{M}(\mathfrak{M}_1 \cup \mathfrak{M}_2 \cup \ldots)$.

PROOF: Since $E_\alpha \leqq E_{\alpha+1}$, Lemma 8 implies that $E_\alpha \cdot E_{\alpha+1} = E_\alpha$. Lemma 6 implies that $E_{\alpha+1}-E_\alpha$ is a projection. Since $E_n = E_1 + \Sigma_{\alpha=1}^{n-1}(E_{\alpha+1}-E_\alpha)$ is a projection, Lemma 5 implies that E_1,

E_2-E_1, E_3-E_2, ... are mutually orthogonal. Furthermore $E = \lim_{n\to\infty} E_n = E_1 + \Sigma_{\alpha=1}^{\infty}(E_{\alpha+1}-E_\alpha)$. Thus Lemma 7 states E is a projection.

The expression for E_n in the previous paragraph and the resulting orthogonality relations, shows that if $\beta \geq n$, $E_n(E_{\beta+1}-E_\beta) = 0$. It follows that $E_nE = E_1 + \Sigma_{\alpha=1}^{n}(E_{\alpha+1}-E_\alpha) = E_n$ and thus by Lemma 8 that $E_n \leq E$.

It follows from Lemmas 7 and 6 that the range of E is $\mathfrak{M}(\mathfrak{M}_1 \cup \mathfrak{M}_1^{\perp}\mathfrak{M}_2 \cup \mathfrak{M}_2^{\perp}\mathfrak{M}_3 \cup \ldots)$. Corollary 1 to Theorem VI of Chapter II, §5 shows that $\mathfrak{U}(\mathfrak{M}_\alpha, \mathfrak{M}_\alpha^{\perp}\cdot\mathfrak{M}_{\alpha+1}) = \mathfrak{M}_{\alpha+1}$. These two results imply that the range of E is $\mathfrak{M}(\mathfrak{M}_1 \cup \mathfrak{M}_2 \cup \ldots)$.

LEMMA 10. Let E_1, E_2, ... be a sequence of projections with ranges $\mathfrak{M}_1$, $\mathfrak{M}_2$, ... respectively and such that $E_\alpha \geq E_{\alpha+1}$. Then $E = \lim_{n\to\infty} E_n$ is a projection such that $E \leq E_\alpha$ for every α, and the range of E is $\mathfrak{M}_1 \cdot \mathfrak{M}_2 \cdot \ldots$.

PROOF: Let $F_\alpha = 1-E_\alpha$. Then $E_\alpha \geq E_{\alpha+1}$, implies $F_\alpha \leq F_{\alpha+1}$, since $F_\alpha \cdot F_{\alpha+1} = (1-E_\alpha)(1-E_{\alpha+1}) = 1-E_\alpha-E_{\alpha+1}+E_\alpha E_{\alpha+1} = 1-E_\alpha = F_\alpha$, when we use Lemma 8 (b). Lemma 9 tells us that $\lim F_\alpha$ is a projection. Hence $1-\lim F_\alpha = \lim(1-F_\alpha) = \lim E_\alpha = E$ is also a projection. Since $\lim F_\alpha \geq F_n$, we must have $E \leq E_n$.

If f is in $\mathfrak{M}_1 \cdot \mathfrak{M}_2 \cdot \ldots$, then $E_\alpha f = f$ for every α and hence $Ef = f$. Thus f is in the range of E and $\mathfrak{M}_1 \cdot \mathfrak{M}_2 \cdot \ldots$ is included in this range. On the other hand, if f is in the range of E, we have $f = Ef$ and since we also have $E_nE = E$ we have $E_nf = E_nEf = Ef = f$. Thus $f \in \mathfrak{M}_n$. Since this holds for every n, $f \in \mathfrak{M}_1 \cdot \mathfrak{M}_2 \cdot \ldots$ and this set includes the range of E. These results together imply the last statement of our lemma.

§2

DEFINITION 1. A transformation $\mathfrak{U}$ from $\mathfrak{H}_1$ to $\mathfrak{H}_2$ with domain $\mathfrak{H}_1$ and range $\mathfrak{H}_2$ and such that $(\mathfrak{U}f, \mathfrak{U}g) = (f,g)$ for every pair of elements in $\mathfrak{H}_1$ is called a unitary transformation from $\mathfrak{H}_1$ to $\mathfrak{H}_2$.

LEMMA 1. A unitary transformation $\mathfrak{U}$ is linear, $\mathfrak{U}^{-1}$ exists and $\mathfrak{U}^* = \mathfrak{U}^{-1}$.

PROOF: Given f_1 and f_2, we have for every g,

$$(\mathfrak{U}(af_1+bf_2)-a\mathfrak{U}f_1-b\mathfrak{U}f_2,\mathfrak{U}g) = (\mathfrak{U}(af_1+bf_2),\mathfrak{U}g)-a(\mathfrak{U}f,\mathfrak{U}g)-b(\mathfrak{U}f_2,\mathfrak{U}g)$$
$$= (af_1+bf_2,g)-a(f_1,g)-b(f_2,g) = 0.$$

Since the set of $\mathfrak{U}g$'s fill out $\mathfrak{H}_2$, we have $\mathfrak{U}(af_1+bf_2) = a\mathfrak{U}f_1+b\mathfrak{U}f_2$ and $\mathfrak{U}$ is additive. Since $|f|^2 = (f,f) = (\mathfrak{U}f,\mathfrak{U}f) = |\mathfrak{U}f|^2$ we see that $\mathfrak{U}$ has bound 1. Thus $\mathfrak{U}$ is linear. Since $|f|^2 = |\mathfrak{U}f|^2$, $\mathfrak{U}f = \theta_2$ implies $f = \theta_1$. Thus $\mathfrak{U}^{-1}$ exists. It is readily seen to be unitary.

Since $\mathfrak{U}$ is linear, $\mathfrak{U}^*$ exists. For every $f \in \mathfrak{H}_1$, $g \in \mathfrak{H}_2$ we have

$$(\mathfrak{U}f,g) = (f,\mathfrak{U}^{-1}g).$$

Thus $\mathfrak{U}^{-1} \subset \mathfrak{U}^*$ and since the former has domain $\mathfrak{H}_2$, $\mathfrak{U}^* = \mathfrak{U}^{-1}$.

LEMMA 2. If $\mathfrak{U}$ is a unitary transformation from $\mathfrak{H}_1$ to $\mathfrak{H}_2$ and $\phi_1, \phi_2, \ldots$ is a complete orthonormal set in $\mathfrak{H}_1$, then $\mathfrak{U}\phi_1, \mathfrak{U}\phi_2, \ldots$ is a complete orthonormal set in $\mathfrak{H}_2$.

PROOF: It is easily seen that the $\mathfrak{U}\phi_1, \mathfrak{U}\phi_2, \ldots$ is an orthonormal set. If g is in $\mathfrak{H}_2$, $g = \mathfrak{U}f$ for an $f \in \mathfrak{H}_1$. $f = \Sigma_{\alpha=1}^{\infty} a_\alpha\phi_\alpha$ by Theorem XII of Chapter II. Thus $g = \mathfrak{U}f = \Sigma_{\alpha=1}^{\infty} a_\alpha\mathfrak{U}\phi_\alpha$ and $\mathfrak{M}(\{\mathfrak{U}\phi_\alpha\}) = \mathfrak{H}_2$. Thus the set $\mathfrak{U}\phi_1, \mathfrak{U}\phi_2, \ldots$ is complete.

LEMMA 3. If $\phi_1, \phi_2, \ldots$ is a complete orthonormal set in $\mathfrak{H}_1$ and $\psi_1, \psi_2, \ldots$ is a complete orthonormal set in $\mathfrak{H}_2$ then the transformation defined by the equation $\mathfrak{U}(\Sigma_{\alpha=1}^{\infty} a_\alpha\phi_\alpha) = \Sigma_{\alpha=1}^{\infty} a_\alpha\psi_\alpha$ is unitary.

This is an immediate consequence of Theorem XII of Chapter II.

DEFINITION 2. An additive transformation V with the property that $(Vf,Vg) = (f,g)$ for every f and g in its domain is called isometric.

LEMMA 4. An isometric transformation V is bounded and has an isometric inverse. $[V]$ exists and is isometric with domain the closure of the domain of V, and range, the closure of the range of V.

PROOF: The first statement is proved in a manner analogous to the proof of the corresponding statements in Lemma 1. Theorem II of §3, of Chapter II implies that $[V]$ exists and has domain the closure of the domain of V. The continuity of the inner product shows that $[V]$ is isometric. A similar argument will show that $[V^{-1}]$ exists and has domain the closure of the range of V. From graphical considerations we see that $[V]^{-1} = [V^{-1}]$. Hence the range of $[V]$ is the closure of the range of V.

LEMMA 5. An additive transformation V, with the property that for every f in its domain $|Vf| = |f|$, is isometric.

This is a consequence of the identity

$$(f,g) = \tfrac{1}{4}(|f+g|^2-|f-g|^2)+\tfrac{1}{4}i(|f+ig|^2-|f-ig|^2).$$

LEMMA 6. Let V be a closed isometric transformation with domain $\mathfrak{M}$ and range $\mathfrak{N}$. $\mathfrak{M}$ and $\mathfrak{N}$ are closed. Let S_1 be an orthonormal set $\phi_1, \phi_2, \ldots$ such that $\mathfrak{M}(S_1) = \mathfrak{M}$. Then $V\phi_1, V\phi_2, \ldots$ is an orthonormal set S_2 such that $\mathfrak{M}(S_2) = \mathfrak{N}$.

$\mathfrak{M}$ and $\mathfrak{N}$ are closed since V is closed and bounded. The proof of the remainder of the Lemma is analogous to the proof of Lemma 2 above.

Theorem XI of Chapter II, §6 has the consequence:

LEMMA 7. Let S_1 denote the orthonormal set $\phi_1, \phi_2, \ldots$ S_2 denote the orthonormal set $\psi_1, \psi_2, \ldots$ and suppose that S_1 and S_2 have the same number of elements. Then the transformation V defined by the equation $V(\Sigma_\alpha a_\alpha \phi_\alpha) = \Sigma_\alpha a_\alpha \psi_\alpha$ is a closed isometric transformation with domain $\mathfrak{M}(S_1)$ and range $\mathfrak{M}(S_2)$.

DEFINITION 3. Let $\mathfrak{M}_1$, $\mathfrak{M}_2$, ... be a sequence of mutually orthogonal manifolds and V_1, V_2, ... be a sequence of isometric transformations such that V_α has domain $\mathfrak{M}_\alpha$ and range $\mathfrak{N}_\alpha$. Let us suppose further that the $\mathfrak{N}_\alpha$'s are mutually orthogonal. It follows from the last paragraph of the proof of Lemma 7, of the preceding section that $\mathfrak{M}(\mathfrak{M}_1 \cup \mathfrak{M}_2 \cup \ldots)$ is the set of elements in the form $\Sigma_\alpha f_\alpha$, $f_\alpha \in \mathfrak{M}_\alpha$. We define

$$V_1 \oplus V_2 \oplus \ldots (\Sigma_\alpha f_\alpha) = \Sigma_\alpha V_\alpha f_\alpha.$$

LEMMA 8. $V_1 \oplus V_2 \oplus \ldots$ is isometric with domain $\mathfrak{M}(\mathfrak{M}_1 \cup \mathfrak{M}_2 \cup \ldots)$ and range $\mathfrak{M}(\mathfrak{N}_1 \cup \mathfrak{N}_2 \cup \ldots)$.

If V_1 and V_2 are closed isometric and V_2 is a proper extension of V_1, the domain and range of V_2 include properly the domain and range of V_1 respectively. Thus $\mathfrak{D}^\perp$ and $\mathfrak{R}^\perp$ for V_1 are not $\{\theta\}$. On the other hand, if $\mathfrak{D}^\perp$ and $\mathfrak{R}^\perp$ are not θ let ϕ' with $|\phi'| = 1$, be $\in \mathfrak{D}^\perp$ and ψ' with $|\psi'| = 1$, be $\in \mathfrak{R}^\perp$. Let V_0 be defined by the equation $V_0(a\phi') = a\psi'$. Then by Lemma 8, $V_1 \oplus V_0$ is a proper isometric extension of V_1. Hence

LEMMA 9. A closed isometric transformation V_1 has a proper isometric extension if and only if both $\mathfrak{D}^\perp$ and $\mathfrak{R}^\perp$ are not $\{\theta\}$.

The dimensionality of a closed linear manifold $\mathfrak{M}$ is the number of elements in an orthonormal set S_1 such that $\mathfrak{M}(S_1) = \mathfrak{M}$. In terms of this definition, we may state

LEMMA 10. A closed isometric transformation V_1 has a unitary extension $\mathfrak{U}$ if and only if the dimensionality of $\mathfrak{D}^\perp$ is the same as that of $\mathfrak{R}^\perp$.

To show that this is necessary, suppose that V has a unitary extension $\mathfrak{U}$. We take a complete orthonormal set $\phi_1, \phi_2, \ldots,$ $\phi_1', \phi_2', \ldots$ with $\phi_1, \phi_2, \ldots \in \mathfrak{D}$, $\phi_1', \phi_2', \ldots \in \mathfrak{D}^\perp$. This can be done by using Theorem VI of §5 of Chapter II and Theorem XI of §6 of Chapter II, because these two results imply that an orthonormal set which consists of a complete orthonormal for $\mathfrak{D}$ and another for $\mathfrak{D}^\perp$ is complete. Since $\mathfrak{U}$ is an extension of V, $\mathfrak{U}\phi_\alpha = V\phi_\alpha \in \mathfrak{R}$ and since $\mathfrak{U}$ is unitary $\mathfrak{U}\phi'_\alpha \in \mathfrak{R}^\perp$. Since the set $\mathfrak{U}\phi_1, \mathfrak{U}\phi_2, \ldots ; \mathfrak{U}\phi'_1, \mathfrak{U}\phi'_2, \ldots$ is complete by Lemma 2 above and $\{\mathfrak{U}\phi_1, \mathfrak{U}\phi_2, \ldots\} = \{V\phi_1, V\phi_2, \ldots\}$ determines $\mathfrak{R}$ by Lemma 6 above, we must have that $\mathfrak{U}\phi'_1, \mathfrak{U}\phi'_2, \ldots$ determines $\mathfrak{R}^\perp$. Thus $\mathfrak{D}^\perp$ and $\mathfrak{R}^\perp$ must have the same dimensionality.

On the other hand, if $\mathfrak{D}^\perp$ and $\mathfrak{R}^\perp$ have the same dimensionality, we can find a partially isometric, V_0 such that $V_0\mathfrak{D}^\perp = \mathfrak{R}^\perp$ by Lemma 7 above. Lemma 8 can be used to show that $V_1 \oplus V_0$ is a unitary extension of V_1.

§3

DEFINITION 1. An additive transformation W from $\mathfrak{H}_1$ to $\mathfrak{H}_2$ which is isometric on a linear manifold $\mathfrak{M}$ and zero on $\mathfrak{M}^\perp$ is called a partially isometric transformation. $\mathfrak{M}$ is called the initial set of W; $\mathfrak{N}$ the range of W is called the final set of W.

LEMMA 1. A partial isometric W is linear. The final set of W is a linear manifold. Let V be the contraction of W with domain $\mathfrak{M}$, let E be the projection of $\mathfrak{H}_1$ on $\mathfrak{M}$. Let F be the projection of $\mathfrak{H}_2$ on $\mathfrak{N}$. Then

$$W = VE = FVE, \quad W^* = V^{-1}F = EV^{-1}F,$$
$$W^*W = E, \quad WW^* = F.$$

PROOF: Since $f = f_1+f_2$, $f_1 \in \mathfrak{M}$, $f_2 \in \mathfrak{M}^\perp$, Wf is defined and equals Wf_1. Hence W has $\mathfrak{H}_1$ as its domain. Since $|Wf_1| = |f_1| \leq |f|$, W is bounded. Thus W is linear.

Since V is bounded and has a closed linear manifold as its domain, V is closed. Thus the range of W which is the range of V is closed. Since $Wf = Wf_1 = Vf_1 = VEf$ for every f, we have $W = VE$. Since the range of V is $\mathfrak{N}$, $W = FVE$. Inasmuch as V^{-1} is isometric on $\mathfrak{N}$, we have

$$(Wf,g) = (FVEf,g) = (VEf,Fg) = (Ef,V^{-1}Fg) = (f,EV^{-1}Fg)$$

for every f and g. Hence $EV^{-1}F \subset W^*$. Since the former is everywhere defined, $W^* = EV^{-1}F$. Now

$$W^*W = EV^{-1}FFVE = EV^{-1}FVE = EV^{-1}VE = E^2 = E.$$

Similarly $WW^* = F$.

LEMMA 2. A c.a.d.d. W such that $W^*W = E$ is a projection is partially isometric with initial set, the range of E.

PROOF: Since W^*W is everywhere defined, W is everywhere defined. If f_1 and g_1 are in the range of E, $Ef_1 = f_1$ and we have $(f_1, g_1) = (Ef_1, g_1) = (W^*Wf, g_1) = (Wf_1, Wg_1)$.

Thus W is isometric on the range of E. If f is orthogonal to this range, $Ef = 0$ and we have

$$0 = (Ef, f) = (W^*Wf, f) = (Wf, Wf) = |Wf|^2$$

and thus $Wf = \theta$. Hence W is partially isometric.

LEMMA 3. The following statements are equivalent for c.a.d.d. transformations: (a) W is isometric, (b) W^* is isometric, (c) W^*W is a projection, (d) WW^* is a projection, (e) $WW^*W = W$, (f) $W^*WW^* = W^*$.

Lemmas 1 and 2 imply that (a) and (c) are equivalent. Similarly (b) and (d) are equivalent. But Lemma 1 shows that (b) implies (a) and also that (a) implies (d). These results show that the first four statements are equivalent.

If W is isometric, we know that $WE = W$ and that $E = W^*W$. Thus $WW^*W = W$ and (a) implies (e). On the other hand, if $WW^*W = W$, we have $W^*WW^*W = W^*W$ or $(W^*W)^2 = W^*W$. Since W^*W is self-adjoint and $(W^*W)^2 = W^*W$, W^*W is a projection by Lemma 2 of §1 of this chapter. Thus (e) implies (c) and hence (e) also must be equivalent to any one of the first four. Taking adjoints shows that (e) is equivalent to (f).

LEMMA 4. Suppose W_α, $\alpha = 1, 2, \ldots$ is a partially isometric transformation with domain $\mathfrak{M}_\alpha$ and range $\mathfrak{N}_\alpha$. We shall suppose that the $\mathfrak{M}_\alpha$'s are mutually orthogonal and that the $\mathfrak{N}_\alpha$'s are mutually orthogonal too. Then $W = \Sigma_\alpha W_\alpha$ is partially isometric with the initial set $\mathfrak{M}(\mathfrak{M}_1 \cup \mathfrak{M}_2 \cup \ldots)$ and the final set $\mathfrak{M}(\mathfrak{N}_1 \cup \mathfrak{N}_2 \cup \ldots)$.

Lemma 8 of the preceding section shows that $\Sigma_\alpha W_\alpha$ is isometric on $\mathfrak{M}(\mathfrak{M}_1 \cup \mathfrak{M}_2 \cup \ldots)$. Now if E_α is the projection on $\mathfrak{M}_\alpha$, $E = \Sigma_\alpha E_\alpha$ is the projection on $\mathfrak{M}(\mathfrak{M}_1 \cup \mathfrak{M}_2 \cup \ldots)$ (Cf. Lemma 7 of §1 of this Chapter). Then $WE = (\Sigma_\alpha W_\alpha)E = \Sigma_\alpha W_\alpha E = \Sigma_\alpha W_\alpha E_\alpha E = \Sigma_\alpha W_\alpha E_\alpha = \Sigma_\alpha W_\alpha = W$. Since $WE = W$, W is zero on the set $\mathfrak{M}(\mathfrak{M}_1 \cup \mathfrak{M}_2 \cup \ldots)^\perp$ and W is isometric.

§4

We return briefly to the considerations of §4 of Chapter IV.

THEOREM I. If A and B are c.a.d.d. operators from $\mathfrak{H}_1$ to $\mathfrak{H}_2$ such that $A^*A = B^*B$, then there exists a partially isometric W with initial set, the closure of the range of A and final set the closure of the range of B such that $B = WA$, $W^*B = A$, $B^* = A^*W^*$, $B^*W = A^*$.

If f is in the domain of A^*A, $Af = \theta$ implies

$$0 = (A^*Af,f) = (B^*Bf,f) = (Bf,Bf) = |Bf|^2$$

or $|Bf|^2 = 0$. Thus the set in $\mathfrak{H}_2 \oplus \mathfrak{H}_2$ of pairs $\{Af,Bf\}$, f in the domain of A^*A, is the graph of a transformation V. The domain of V is the range of A' the contraction of A with domain the domain of A^*A.

Now V is isometric since for f and g in the domain of A^*A,

$$(Af,Ag) = (A^*Af,g) = (B^*Bf,g) = (Bf,Bg).$$

Thus if $\phi = Af$, $\psi = Ag$, $V\phi = Bf$, $V\psi = Bg$ and

$$(\phi,\psi) = (V\phi,V\psi)$$

for every ϕ and ψ in the domain of V. Furthermore $VA' = B'$ where B' is the contraction of B with domain the domain of B^*B.

Let f be an element in the domain of $[A']$. We can find a sequence $\{f_n,A'f_n\}$ such that $\{f_nA'f_n\} \longrightarrow \{f,[A']f\}$. Hence the sequence $\{A'f_n\}$ converges and owing to the isometry relation, V, the sequence $\{B'f_n\}$ must converge to a g^*. Thus $\{f_n,B'f_n\}$ converges to $\{f,g^*\}$. Then by the definition of $[B']$, $[B']f$ exists and equals g^*. We also have $\{A'f_n,B'f_n\} \longrightarrow \{[A']f,[B']f\}$. Hence this latter pair is in the graph of

$[V]$ and $[V][A']f = [B']f$. This last equation holds for every f in the domain of $[A']$ and hence $[V][A'] \subset [B']$. If we take f in the domain of $[B']$, a precisely similar argument shows the reverse inclusion and hence $[V][A'] = [B']$.

Theorem VIII of §4 of Chapter IV states that $[A'] = A$, $[B'] = B$. Thus $[V]A = B$. It is obvious from the graphical considerations made in the above that $A = [V]^{-1}B$, that the domain of $[V]$ is the closure of the range of A and that the range of $[V]$ is the closure of the range of B. Let the closure of the range of A have a projection E and that of B have a projection F. Let $W = [V]E$. Then $W = F[V]E$, $W^* = E[V]^{-1}F$, W is partially isometric with initial set the range of E, and final set the range of F. $B = [V]A = [V]EA = WA$. $A = [V]^{-1}B = E[V]^{-1}FB = W^*B$.

Now if $B = WA$, $B^* = A^*W^*$ by the Corollary to Theorem V of §2, Chapter IV. Similarly $A^* = B^*W$.

It should be remarked that the partially isometric W is introduced because in general $[V]^*$ does not exist.

CHAPTER VII

RESOLUTIONS OF THE IDENTITY

§1

In this section we will discuss certain properties of self-adjoint transformations whose range is finite dimensional. While these results will be applied later, they should also be regarded as indicating what results are desired in the general case.

LEMMA 1. Let H be a self-adjoint transformation with a finite dimensional range $\mathfrak{M}$ which is determined by the orthonormal set $\phi_1, \ldots, \phi_n$. Then H is zero in $\mathfrak{M}^\perp$ and there is an n'th order matrix $(a_{\alpha,\beta})$ $\alpha, \beta = 1, \ldots, n$ with $a_{\alpha,\beta} = \bar{a}_{\beta,\alpha}$ such that $H\phi_\alpha = \Sigma_\beta a_{\alpha,\beta}\phi_\beta$. H is bounded.

Since $H = H^*$, H is zero on $\mathfrak{M}^\perp$ by Lemma 8 of §3 of Chapter IV. Let E be the projection on $\mathfrak{M}$. $\mathfrak{M}^\perp \subset \mathfrak{D}_H$ implies that for $f \in \mathfrak{D}_H$, $(1-E)f \in \mathfrak{D}_H$ and $Ef \in \mathfrak{D}_H$. Since $\mathfrak{D}_H$ is dense, this implies that $\mathfrak{D}_H$ is dense in $\mathfrak{M}$. Since $\mathfrak{M}$ is finite dimensional and $\mathfrak{D}_H$ is additive, $\mathfrak{D}_H$ must contain $\mathfrak{M}$. Since $H\phi \in \mathfrak{M}$, $H\phi = \Sigma_{\beta=1}^{n} a_{\alpha,\beta}\phi_\beta$ by Theorem XI of Chapter II, §6. $a_{\alpha,\beta} = (H\phi_\alpha, \phi_\beta) = (\phi_\alpha, H\phi_\beta) = \overline{(H\phi_\beta, \phi_\alpha)} = \bar{a}_{\beta,\alpha}$.

The bound of H is the bound of its contraction defined on $\mathfrak{M}$ and for this we have

$$\begin{aligned} |H\Sigma_\alpha x_\alpha\phi_\alpha|^2 &= |\Sigma_\alpha x_\alpha H\phi_\alpha|^2 \\ = |\Sigma_\beta(\Sigma_\alpha x_\alpha a_{\alpha,\beta})\phi_\beta|^2 &= \Sigma_\beta|\Sigma_\alpha x_\alpha a_{\alpha,\beta}|^2 \\ &\leq (\Sigma_{\alpha,\beta}|a_{\alpha,\beta}|^2)(\Sigma_\alpha|x_\alpha|^2). \end{aligned}$$

The converse of Lemma 1 holds.

LEMMA 2. A finite orthonormal set S, $\phi_1, \ldots, \phi_n$, and a symmetric matrix $(a_{\alpha,\beta})$ $\alpha,\beta = 1, \ldots, n$ (i.e., $a_{\alpha,\beta} = \bar{a}_{\beta,\alpha}$) determine a self-adjoint transformation by means of the conditions: If $f \in \mathfrak{M}(S)^\perp$, $Hf = \theta$; $H\phi_\alpha = \Sigma_\beta a_{\alpha,\beta}\phi_\beta$.

H is readily seen to be symmetric and defined everywhere.

The essential result of this section is that $\phi_1, \ldots, \phi_n$ can be chosen so that $a_{\alpha,\beta} = \lambda_\alpha \delta_{\alpha,\beta}$ for real λ_α. Thus we obtain the "diagonal form" for the matrix.

LEMMA 3. If H is a non-zero self-adjoint transformation whose range is a finite dimensional $\mathfrak{M}$, then we can find a ϕ in $\mathfrak{M}$ and a non-zero λ such that if $\mathfrak{M}_1$ is the set of ψ in $\mathfrak{M}$ which are orthogonal to ϕ, then for every f in $\mathfrak{H}$,

$$Hf = \lambda(f,\phi)\phi + H_1 f$$

where H_1 is a self-adjoint transformation whose range is $\mathfrak{M}_1$.

PROOF: Since H is not zero, either $C_+ > 0$ or $C_- < 0$. (Cf. Definition 4 of §3 of Chapter IV). We shall suppose $C_+ > 0$. (Otherwise our argument would apply to $-H$) Let E' be the projection on $\mathfrak{M}$. Then $EHE = H$ since H is zero on $\mathfrak{M}^\perp$. For every f we have $(Hf,f) = (EHEf,f) = (HEf,Ef)$.

Now let $f_1, f_2, \ldots$ be a sequence of elements with $|f_n| = 1$ and such that $(Hf_n, f_n) \longrightarrow C_+$. It follows that $|Ef_n| \leqq 1$, $(HEf_n, Ef_n) \longrightarrow C_+$. All the Ef_n 's are in $\mathfrak{M}$ whose unit sphere is compact. Thus a subsequence of the Ef_n 's must converge to a g in $\mathfrak{M}$ such that $(Hg,g) = C_+$, $|g| \leqq 1$. Furthermore $|g| = 1$, since if $|g| < 1$, $(Hg,g)/|g|^2 > C_+$ a contradiction.

We let $\phi = g$. If $\psi \in \mathfrak{M}_1$, then $(\psi,\phi) = 0$ and for every value of α, $|\cos\alpha\phi + \sin\alpha\psi|^2 = 1$. Thus if $f = \cos\alpha\phi + \sin\alpha\psi$,

$$(H\phi,\phi) = C_+ \geqq (Hf,f) = \cos^2\alpha(H\phi,\phi) + 2\sin\alpha\cos\alpha R(H\phi,\psi) + \sin^2\alpha(H\psi,\psi).$$

This is only possible if $R(H\phi,\psi) = 0$. Multiplying ψ by a constant does not effect $\psi \in \{\phi\}^\perp$ and thus we have by a familiar process $|(H\phi,\psi)| = 0$.

Let $\phi, \psi_1, \ldots, \psi_{n-1}$ be an orthonormal set complete in $\mathfrak{M}$ and thus

$$H\phi = \lambda\phi + b_1\psi_1 + \ldots + b_{n-1}\psi_{n-1}.$$

But ψ_α is in $\mathfrak{M}_1$ and thus $b_\alpha = (H\phi,\psi_\alpha) = 0$. Hence $H\phi = \lambda\phi$, where $\lambda = (H\phi,\phi) = C_+$ is real and not zero.

If E_1 is the projection with range $\mathfrak{M}_1$, we have

$(HE_1f,\phi) = (E_1f,H\phi) = 0$ for every f. Thus the range of HE_1 is orthogonal to ϕ. The range of HE_1 is included in $\mathfrak{M}$ and this means that it is included in $\mathfrak{M}_1$. Hence $HE_1 = E_1HE_1$. E_1HE_1 is self-adjoint.

Now the projection on $\{\alpha\phi\}$ is given by the equation $E_\phi f = (f,\phi)\phi$. Hence $Ef = (f,\phi)\phi+E_1f$ and

$$Hf = HEf = (f,\phi)H\phi+HE_1f = \lambda(f,\phi)\phi+H_1f$$

where $H_1 = HE_1 = E_1HE_1$ is self adjoint with range included in $\mathfrak{M}_1$. Dimensionality considerations show that the range of H_1 must be $\mathfrak{M}_1$.

Call ϕ, ϕ_1 and apply Lemma 3 to H_1. This gives us a ϕ_2 in $\mathfrak{M}_1$ such that

$$Hf = \lambda_1(f,\phi_1)\phi_1+\lambda_2(f,\phi_2)+H_2f$$

where H_2 has range in $\mathfrak{M}_1$ orthogonal to ϕ_2 and of course to ϕ_1. Repeating, we obtain the following lemma.

LEMMA 4. If H is a self-adjoint transformation with a finite dimensional range $\mathfrak{M}$, we can find an orthonormal set S, $\phi_1, \ldots, \phi_n$ and real non-zero constants $\lambda_1, \ldots, \lambda_n$ such that

$$Hf = \lambda_1(f,\phi_1)\phi_1+\lambda_2(f,\phi_2)\phi_2+ \ldots +\lambda_n(f,\phi_n)\phi_n.$$

The λ_α satisfy the inequality $C_- \leqq \lambda_\alpha \leqq C_+$.

For such a transformation, H^2, H^3, ... can be defined. Let $p(x) = a_nx^n+a_{n-1}x^{n-1}+ \ldots +a_0$ be a polynomial. We define $p'(H) = a_nH^n+a_{n-1}H^{n-1}+ \ldots +a_0E$ where E is the projection on $\mathfrak{M}$. Then it is easily verified that

$$p'(H) = p(\lambda_1)(f,\phi_1)\phi_1+p(\lambda_2)(f,\phi_2)\phi_2+ \ldots +p(\lambda_n)(f,\phi_n)\phi_n.$$

In general we have

LEMMA 5. The correspondence $p(x) \backsim p'(H)$ preserves the operation of addition, multiplication and multiplication by a constant.

$$p'(H) = a_nH^n+a_{n-1}H^{n-1}+ \ldots +a_0E$$

$$= p(\lambda_1)(f,\phi_1)\phi_1+ \ldots + p(\lambda_n)(f,\phi_n)\phi_n$$

when $p(x) = a_n x^n + a_{n-1} x^{n-1} + \dots + a_0$. When the a 's are real and C_1 and C_2 are two constants such that for $C_- \leq x \leq C_+$, we have $C_1 \leq p(x) \leq C_2$ then for every f,

$$C_1(Ef,Ef) \leq (p'(H)f,f) \leq C_2(Ef,Ef).$$

To show the last statement we note that $p'(H) = Ep'(H)$. Thus

$$(p'(H)f,f) = (p'(H)f,Ef)$$

$$= (p(\lambda_1)(f,\phi_1)\phi_1 + \dots + p(\lambda_n)(f,\phi_n)\phi_n, (f,\phi_1)\phi_1 + \dots + (f,\phi_n)\phi_n)$$

$$= p(\lambda_1)|(f,\phi_1)|^2 + \dots + p(\lambda_n)|(f,\phi_n)|^2.$$

This will imply the desired inequality.

We will obtain an infinite analogue of Lemmas 4 and 5. However it must be remembered that in dealing with an infinite dimensional space, one must consider not sums but limits of sums. Thus $\Sigma_{\alpha=1}^{\infty}$ (or $\Sigma_{\alpha=-\infty}^{\infty}$) represents that special limiting process in which one (or both) limits of summation are permitted to approach ∞. The integral $\int$ in the Rieman-Stieljes sense, is a more general process of taking the limit of a sum, which includes the preceding method. Thus the generalization of the expression for Hf in Lemma 4 need not be an infinite sum

$$\Sigma_{\alpha=-\infty}^{\infty} \lambda_\alpha (f,\phi_\alpha)\phi_\alpha$$

but a more general method of taking the limit of a sum.

§2

DEFINITION 1. A family of projections $E(\lambda)$ defined for $-\infty < \lambda < +\infty$ is called a resolution of the identity if

1. $E(\lambda) \geq E(\mu)$ for $\lambda > \mu$.
2. $E(\lambda+0) = E(\lambda)$. *
3. $\lim_{\lambda \to -\infty} E(\lambda) = 0$, $\lim_{\lambda \to \infty} E(\lambda) = 1$.

A resolution of the identity will be said to be finite, if there is a λ_1 such that $E(\lambda_1) = 0$ and a λ_2 such that $E(\lambda_2) = 1$. **

* $E(\lambda+0) = \lim_{\epsilon \to 0} E(\lambda+\epsilon^2)$, $E(\lambda-0) = \lim_{\epsilon \to 0} E(\lambda-\epsilon^2)$.

**The following are examples of a resolution of the identity.
(a) Let $\dots \phi_{-1}, \phi_0, \phi_1, \dots$ be a complete orthonormal set

It follows from Lemma 6 and 8 of Chapter VI, §1, that if $\lambda_1 > \lambda_2$, $E(\lambda_1) - E(\lambda_2)$ is a projection. It follows from Lemma 5 of Chapter VI §1, that if $\lambda_1 > \lambda_2 \geq \mu_1 > \mu_2$, then $(E(\lambda_1)-E(\lambda_2))(E(\mu_1)-E(\mu_2)) = 0$ since $E(\lambda_1)-E(\mu_2) = E(\lambda_1)-E(\lambda_2)+E(\lambda_2)-E(\mu_1)+E(\mu_1)-E(\mu_2)$. It is also a consequence of Lemma 8 that $E(\lambda_1)E(\lambda_2) = E(\min(\lambda_1,\lambda_2)) = E(\lambda_2)E(\lambda_1)$.

DEFINITION 2. If $b > a$, we shall define a "partition" Π of the interval (a,b) as a set of points $x_0 = a$, x_1, x_2, ... , $x_n = b$, with $x_\alpha < x_{\alpha+1}$ which subdivides the interval (a,b) into n smaller intervals $(x_{\alpha-1},x_\alpha)$. The interval $(x_{\alpha-1},x_\alpha)$ will be said to be marked if a point x'_α with $x_{\alpha-1} \leqq x'_\alpha \leqq x_\alpha$ is chosen in it. If each smaller interval of Π is marked, we will say that the partition Π is marked and denote the marked partition Π'. If Π_0 is a subdivision y_0,y_1, ... , y_m such that every $x_\alpha = y_{\beta_\alpha}$ for some β_α then Π_0 is called a finer subdivision of Π. We indicate this $\Pi_0 \dashv \Pi$. The mesh of a subdivision $m(\Pi) = \max(x_\alpha - x_{\alpha-1})$. If $E(\lambda)$ is a resolution of the identity, $\phi(\lambda)$ a complex valued function defined on the interval $a \leqq \lambda \leqq b$ and Π' a marked subdivision of this interval we define

$$\Sigma_{\Pi'}\phi\Delta E(\lambda) = \Sigma^n_{\alpha=1}\phi(x'_\alpha)(E(x_\alpha)-E(x_{\alpha-1})).$$

LEMMA 1. $\Sigma_{\Pi'}\phi\Delta E(\lambda)$ is a bounded transformation with bound $C = \max|\phi(x'_\alpha)|$. We also have

(a) $\Sigma_{\Pi'}\phi_1\Delta E(\lambda) + \Sigma_{\Pi'}\phi_2\Delta E(\lambda) = \Sigma_{\Pi'}(\phi_1+\phi_2)\Delta E(\lambda)$.

(b) $(\Sigma_{\Pi'}\phi_1\Delta E(\lambda))(\Sigma_{\Pi'}\phi_2\Delta E(\lambda)) = \Sigma_{\Pi'}\phi_1\phi_2\Delta E(\lambda)$.

(c) $(\Sigma_{\Pi'}\phi\Delta E(\lambda)f,f) = \Sigma^n_{\alpha=1}\phi(x'_\alpha)((E(x_\alpha)-E(x_{\alpha-1}))f,f)$.

(d) $|\Sigma_{\Pi'}\phi\Delta E(\lambda)f|^2 = \Sigma^n_{\alpha=1}|\phi(x'_\alpha)|^2\cdot|(E(x_\alpha)-E(x_{\alpha-1}))f|^2$.

---

whose indices range over the integers from $-\infty$ to ∞. If $\mathfrak{M}(\lambda)$ is the manifold determined by the ϕ_α for which $\alpha \leqq \lambda$, and $E(\lambda)$ is the projection on $\mathfrak{M}(\lambda)$, then one can easily verify that $E(\lambda)$ is a resolution of the identity.

(b) Let us realize $\mathfrak{H}$ as $\mathfrak{L}_2$ (Cf. Chapter III, §4, Definition 1). If $\lambda \leqq 0$, we define $\chi_\lambda(x) = 0$, If $0 < \lambda \leqq 1$, $\chi_\lambda(x) = 1$ when $x \leqq \lambda$, $\chi_\lambda(x) = 0$ if $\lambda < x$. If $\lambda > 1$, $\chi_\lambda(x) = 1$. It is readily verified that the transformations defined by the equation $E(\lambda)f(x) = \chi_\lambda(x)f(x)$ form a resolution of the identity.

These results are immediate consequences of

$$(E(\lambda_1)-E(\lambda_2))(E(\mu_1)-E(\mu_2)) = 0 \quad \text{if} \quad \lambda_1 > \lambda_2 \geqq \mu_1 > \mu_2.$$

Thus

$$|\Sigma_{\Pi}\phi\Delta E(\lambda)f|^2 = \Sigma_{\alpha=1}^{m}|\phi(x_\alpha')|^2\cdot|(E(x_\alpha)-E(x_{\alpha-1}))f|^2$$
$$\leqq C^2(\Sigma_{\alpha=1}^{n}|(E(x_\alpha)-E(x_{\alpha-1}))f|^2)$$
$$= C^2|(E(b)-E(a))f|^2 \leqq C^2|f|^2.$$

DEFINITION 3. We shall say that $\int_a^b\phi(\lambda)dE(\lambda)f$ exists for a given f if there is an element f^* in $\mathfrak{H}$ such that for every sequence of partitions $\Pi_1, \Pi_2, \ldots$ such that $\Pi_\alpha < \Pi_{\alpha-1}$ and $m(\Pi_n) \longrightarrow 0$ as $n \longrightarrow \infty$, then

$$|f^*-\Sigma_{\Pi_n'}\phi\Delta E(\lambda)f| \longrightarrow 0.$$

We define $\int_a^b\phi(\lambda)dE(\lambda)f = f^*$.

LEMMA 2. Let $\phi(\lambda)$ be continuous on the interval (a,b), and let Π' and Π_0' denote two marked subdivisions with Π_0 a finer subdivision of Π. Given an $\epsilon > 0$, there is a number $\mu = \mu(\epsilon)$ such that if the mesh of Π, $m(\Pi)$ is $\leqq \mu(\epsilon)$, then

$$|\Sigma_{\Pi'}\phi\Delta E(\lambda)f-\Sigma_{\Pi_0'}\phi\Delta E(\lambda)f| \leqq \epsilon|f|.$$

PROOF: Since ϕ is uniformly continuous, we may define $\mu(\phi)$ as the $\delta(\epsilon) > 0$, such that when $|x_1-x_2| < \delta(\epsilon)$ then $|\phi(x_1)-\phi(x_2)| < \epsilon$. Let us suppose $m(\Pi) \leqq \mu$. We define $\phi_{\Pi'}(x)$ by the equation $\phi_{\Pi'}(x) = \phi(x_\alpha')$ if $x_{\alpha-1} \leqq x < x_\alpha$. Since $m(\Pi) \leqq \mu$, $|\phi_{\Pi'}(x)-\phi(x)|$. Furthermore for $\Pi_0 < \Pi$ $\Sigma_{\Pi_0'}\phi_{\Pi'}\Delta E(\lambda) = \Sigma_{\Pi}\phi\Delta E(\lambda)$.
Hence the transformation

$$\Sigma_{\Pi'}\phi\Delta E(\lambda)-\Sigma_{\Pi_0'}\phi\Delta E(\lambda)$$
$$= \Sigma_{\Pi_0'}\phi_{\Pi'}\Delta E(\lambda)-\Sigma_{\Pi_0'}\phi\Delta E(\lambda)$$
$$= \Sigma_{\Pi_0'}(\phi'-\phi)\Delta E(\lambda)$$

has a bound $\leqq \epsilon$ by Lemma 1 above.

The existence of $\int_a^b\phi(\lambda)dE(\lambda)f$ follows from Lemma 2 in a manner entirely analoguous to the proof of the existence of the

Rieman-Stieljes integral in the ordinary sense. Thus Lemma 2 implies that every sequence of partitions $\Pi_1, \Pi_2, \ldots$ with $\Pi_\alpha < \Pi_{\alpha-1}$ and $m(\Pi_n) \longrightarrow 0$ as $n \longrightarrow \infty$ will have $\Sigma_{\Pi_\alpha}{}' \phi \Delta E(\lambda) f$ convergent. The sequences $\Pi_{1,1}, \Pi_{2,1}, \ldots$ and $\Pi_{1,2}, \Pi_{2,2}, \ldots$ will have the corresponding Σ convergent to the same limit as one easily sees if one considers a subdivision Π_n which is a finer subdivision of both $\Pi_{n,1}$ and $\Pi_{n,2}$.

THEOREM I. If $\phi(x)$ is a continuous function on the interval $a \leq x \leq b$, then for every f,

$$Tf = \int_a^b \phi(\lambda) dE(\lambda) f$$

exists. Tf is linear with bound $\leq \max_{a \leq x \leq b} \phi(x)$. If $\phi(x)$ is real, T is also self-adjoint.

The continuity and the bound of T are consequences of Lemma 1. Suppose $\phi(x)$ is real. If we let for the moment $T_n = \Sigma_{\Pi_n} \phi \Delta E(\lambda)$ we see that each T_n is self-adjoint. Thus for every f and g.

$$(Tf,g) = \lim(T_n f,g) = \lim(f,T_n g) = (f,Tg).$$

Hence T is symmetric and since it is linear it is self-adjoint.

LEMMA 3. If $H = \int_a^b \phi(\lambda) dE(\lambda)$ for $\phi(\lambda)$ real and continuous then

$$(Hf,f) = \int_a^b \phi(\lambda) d(E(\lambda)f,f)$$

where the integral is the Rieman-Stieljes one in the ordinary sense. C_+ for H is $\leq \max_{a \leq x \leq b} \phi(x) = M$, when $M \geq 0$. C_- for H is $\geq \min_{a \leq x \leq b} \phi(\bar{x}) = m$ when $m \leq 0$,

$$|Hf|^2 = \int_a^b |\phi(\lambda)|^2 d|E(\lambda)f|^2.$$

This Lemma follows readily from Lemma 1 above.

We may remark that the conditions $M \geq 0$, $m \leq 0$ in Lemma 3 are necessary because there may be non-zero f 's such that $(E(b)-E(a)f = \theta$. If no such f occurs, these conditions may be omitted. (Cf. Lemma 1 above and its proof.)

LEMMA 4. If $a \leq a' < b' \leq b$

$$(\int_a^b\phi(\lambda)dE(\lambda))(E(b')-E(a'))$$
$$= (E(b')-E(a'))\int_a^b\phi(\lambda)dE(\lambda) = \int_{a'}^{b'}\phi(\lambda)dE(\lambda).$$

This is easily seen if one considers the orthogonality relations for the differences $E(x_\alpha)-E(x_{\alpha-1})$.

We can now point out the precise circumstances under which, we will have $C_+ < \max_{a \leq x \leq b}\phi(x) = M$, $M > 0$ in Lemma 3. Let λ_0 be a point of the interval (a,b) for which $\phi(\lambda_0) = M$. Then there are two numbers λ_1 and λ_2 in the interval (a,b) with $\lambda_1 \leqq \lambda_0 \leqq \lambda_2$, $\lambda_1 \neq \lambda_2$ and such that for $\lambda_1 \leqq \lambda \leqq \lambda_2$, $\phi(\lambda) \geq C_++\delta$ for a $\delta > 0$.

For any such pair, we must have $E(\lambda_2)-E(\lambda_1) = 0$. For let f be in the range of $E(\lambda_2)-E(\lambda_1)$. Then Lemma 8 of Chapter VI, §1 and the preceding results imply

$$(\int_a^b\phi(\lambda)dE(\lambda)f,f) = (\int_a^b\phi(\lambda)dE(\lambda)(E(\lambda_2)-E(\lambda_1))f,f)$$
$$= (\int_{\lambda_1}^{\lambda_2}\phi(\lambda)dE(\lambda)f,f) = \int_{\lambda_1}^{\lambda_2}\phi(\lambda)d(E(\lambda)f,f)$$
$$= \int_{\lambda_1}^{\lambda_2}\phi(\lambda)d|E(\lambda)f|^2 \geq (C_++\delta)\int_{\lambda_1}^{\lambda_2}d|E(\lambda)f|^2$$
$$= (C_++\delta)\int_{\lambda_1}^{\lambda_2}d(E(\lambda)f,f) = (C_++\delta)((E(\lambda_2)-E(\lambda_1))f,f)$$
$$= (C_++\delta)|f|^2.$$

The definition of C_+ implies that this is only possible if $|f|^2 = 0$.

A further consequence of this situation is

$$\int_a^b\phi(\lambda)dE(\lambda) = \int_a^{\lambda_1} \phi(\lambda)dE(\lambda)+\int_{\lambda_2}^b \phi(\lambda)dE(\lambda).$$

LEMMA 5. For $\phi_1(\lambda)$ and $\phi_2(\lambda)$ continuous, we have

(a) $$\int_a^b\phi_1 dE(\lambda)+\int_a^b\phi_2 dE(\lambda) = \int_a^b(\phi_1+\phi_2)dE(\lambda)$$

(b) $$(\int_a^b\phi_1(\lambda)dE(\lambda))(\int_a^b\phi_2(\lambda)dE(\lambda)) = \int_a^b\phi_1\phi_2 dE(\lambda).$$

PROOF: The statement (a) is obvious. To show (b) we note first that since ϕ_1 is continuous on a closed interval $|\phi_1(\lambda)|$ is bounded by some C. As in Lemma 2, we can find a function $\mu(\epsilon)$ defined for $\epsilon > 0$ such that if $|x_1-x_2| < \mu(\epsilon)$ $|\phi_2(x_1)-\phi_2(x_2)| < \epsilon$ and also such that if $m(\Pi) < \mu(\epsilon)$ and $\Pi_0 < \Pi$, then

$$|\Sigma_{\Pi'}\phi_2\Delta E(\lambda)f-\Sigma_{\Pi_0'}\phi_2\Delta E(\lambda)f| < \epsilon|f|.$$

Letting $m(\Pi_0') \longrightarrow 0$, we have

$$|\Sigma_{\Pi'}\phi_2\Delta E(\lambda)f - \int_a^b\phi_2(\lambda)dE(\lambda)f| < \epsilon|f|.$$

Hence by Lemma 3, since $|\phi_1| < C$,

$$|\int_a^b\phi_1(\lambda)dE(\lambda)\,\Sigma_{\Pi'}\phi_2\Delta E(\lambda)f - (\int_a^b\phi_1(\lambda)dE(\lambda))\int_a^b\phi_2 dE(\lambda)f| \leq C\epsilon|f|.$$

Now let $\phi_{2,\Pi'}$ be defined as in Lemma 2, i.e. $\phi_{2,\Pi'}(x) = \phi_2(x_\alpha')$ if $x_{\alpha-1} \leq x < x_\alpha$. Then Lemma 4 implies

$$(\int_a^b\phi_1 dE(\lambda))\Sigma_{\Pi'}\phi_2\Delta E(\lambda)f = \Sigma_{\alpha=1}^n\phi_2'(x_\alpha')\int_{x_{\alpha-1}}^{x_\alpha}\phi_1(\lambda)dE(\lambda)$$
$$= \int_a^b\phi_{2,\Pi'}(\lambda)\phi_1(\lambda)dE(\lambda).$$

The preceding inequality now implies

$$|\int_a^b\phi_{2,\Pi'}\phi_1 dE(\lambda) - (\int_a^b\phi_2 dE(\lambda))(\int_a^b\phi_1 dE(\lambda))| \leq C\epsilon f.$$

However $|\phi_{2,\Pi'}(x)-\phi_2(x)| < \epsilon$. Thus

$$|\int_a^b\phi_{2,\Pi'}\phi_1 dE(\lambda)f - \int_a^b\phi_1\phi_2 dE(\lambda)f|^2 = |\int_a^b\phi_1(\phi_{2,\Pi'}-\phi_2)dE(\lambda)f|^2$$
$$= \int_a^b|\phi_1|^2(\phi_{2,\Pi'}-\phi_2)^2 d|E(\lambda)f|^2 \leq C^2\epsilon^2|f|^2.$$

This and the last inequality of the preceding paragraph imply

$$|(\int_a^b\phi_2 dE(\lambda))(\int_a^b\phi_1 dE(\lambda))f - \int_a^b\phi_1\phi_2 dE(\lambda)f| \leq 2C\epsilon|f|.$$

Since ϵ is arbitrary, we must have for every f,

$$(\int_a^b\phi_2 dE(\lambda))(\int_a^b\phi_1 dE(\lambda)) = \int_a^b\phi_1\phi_2 dE(\lambda),$$

which implies the last statement of our Lemma.

LEMMA 6. If $\phi(x)$ is continuous and > 0 for $a < x \leq b$ then the zeros of $T = \int_a^b\phi(\lambda)dE(\lambda)$ form the range of $1-E(b)+E(a)$.

Let $\mathfrak{N}_0$ denote the range of $1-E(b)+E(a)$. Since every $E(x_\alpha)-E(x_{\alpha-1})$ in $\Sigma_\Pi\phi\Delta E(\lambda)$ is orthogonal to $1-E(b)+E(a)$ it follows that $\mathfrak{N}_0 \subset \mathfrak{N}$. $\mathfrak{N}_0^\perp$ is the range of $E(b)-E(a)$ and we shall show that if $f \neq \theta$ is in $\mathfrak{N}_0^\perp$, then $Tf \neq \theta$.

Since $E(a+0) = E(a)$ (Cf. Definition 1 of this section) and since $(E(b)-E(a))f = f \neq \theta$, we can find an a' such that $b > a' > a$ and $(E(b)-E(a')f \neq \theta$. Since $\phi(x) > 0$ and continuous for $b \geq x \geq a'$, it follows that there is a $\delta > 0$, such that $\phi(x) \geq \delta$ in this interval.

$$(Tf,f) = \int_a^b\phi(x)d(E(\lambda)f,f) \geq \int_{a'}^b\phi(x)d(E(\lambda)f,f)$$
$$\geq \delta((E(b)-E(a'))f,f) = \delta|(E(b)-E(a'))f|^2 > 0.$$

Hence $Tf \neq \theta$ and $\mathfrak{N}_0^{\perp}\cdot\mathfrak{N} = \{\theta\}$. Since $\mathfrak{N}_0 \subset \mathfrak{N}$ Corollary 1 to Theorem VI of Chapter II, §5, will show that $\mathfrak{N}_0 = \mathfrak{N}$.

If $\phi(b) = 0$, the same sort of argument will yield that the zeros of T form the range of $1-E(b-0)+E(a)$. If $\phi(c) = 0$ for a c between a and b and $\phi(x) \neq 0$ otherwise, the zeros of T form the range of $1-E(b)+E(c)-E(c-)+E(a)$. This result is easily generalized to the case of n zeros.

§3

We deal in this section with "improper" integrals. There are a number of possibilities for improper integrals. For instance, we may have that $\phi(x)$ has a singularity at $x = a$. Then we define

$$\int_a^b \phi(\lambda)dE(\lambda)f = \lim_{\epsilon\to 0} \int_{a+\epsilon^2}^b \phi(\lambda)E(\lambda)f$$

when this limit exists. This will be called an improper integral of the first kind. The second kind is defined by the equation,

$$\int_{-\infty}^{\infty} \phi(\lambda)dE(\lambda)f = \lim_{a\to-\infty,\, b\to+\infty} \int_a^b \phi(\lambda)dE(\lambda)f$$

when this limit exists. The discussion that follows is analogous in each case. We will therefore at times prove our results only for the second kind and assume them for both kinds.

LEMMA 1. If $-\infty < a < b < \infty$, then

$$(E(b)-E(a))\int_{-\infty}^{\infty} \phi(\lambda)dE(\lambda) \subset \left(\int_{-\infty}^{\infty} \phi(\lambda)dE(\lambda)\right)(E(b)-E(a)) = \int_a^b \phi(\lambda)dE(\lambda).$$

PROOF: If f is such that $\int_{-\infty}^{\infty} \phi(\lambda)dE(\lambda)f$ is defined, we have by Lemma 4 of the preceding section, that

$$(E(b)-E(a))\int_{-\infty}^{\infty} \phi(\lambda)dE(\lambda)f = (E(b)-E(a)) \lim \int_{a'}^{b'} \phi(\lambda)dE(\lambda)f$$
$$= \lim\, (E(b)-E(a))\int_{a'}^{b'} \phi(\lambda)dE(\lambda)f = \lim \int_a^b \phi(\lambda)dE(\lambda)f = \int_a^b \phi(\lambda)dE(\lambda)f$$
$$= \lim \int_{a'}^{b'} \phi(\lambda)dE(\lambda)(E(b)-E(a))f = \int_{-\infty}^{\infty} \phi(\lambda)dE(\lambda)(E(b)-E(a))f.$$

This proves the inclusion and the equality is proved by a similar argument.

LEMMA 2. For a fixed $f \in \mathfrak{H}$, the following statements are equivalent.

(a) $\int_{-\infty}^{\infty} \phi(\lambda)dE(\lambda)f$ exists.

(b) $\int_{-\infty}^{\infty} |\phi(\lambda)|^2 d|E(\lambda)f|^2 < \infty.$

(c) For every g,

$$F(g) = \int_{-\infty}^{\infty} \phi(\lambda)d(E(\lambda)g,f)$$

exists and $F(g)$ is a linear functional.

We first show that (c) implies (b). By Theorem I of §3, Chapter II, there exists a constant C such that $|F(g)| \leq C|g|$. Let $g = \int_a^b \phi(\lambda)dE(\lambda)f$. Lemma 4 of the preceding section implies that $(E(b)-E(a))g = g$ and thus

$$\overline{F}(g) = \int_{-\infty}^{\infty} \phi(\lambda)d(E(\lambda)f,g) = \lim \int_{a'}^{b'} \phi(\lambda)d(E(\lambda)f,g)$$

$$= \lim (\int_{a'}^{b'} \phi(\lambda)d(E(\lambda)f,g)) = \lim (\int_{a'}^{b'} \phi(\lambda)dE(\lambda)f,(E(b)-E(a))g)$$

$$= \lim ((E(b)-E(a))\int_{a'}^{b'} \phi(\lambda)dE(\lambda)f,g) = (\int_a^b \phi(\lambda)dE(\lambda)f,g)$$

$$= (g,g) = |g|^2.$$

Thus $|g|^2 \leq C\cdot|g|$ or $|g| \leq C$. Lemma 3 of the preceding section now implies

$$\int_a^b |\phi(\lambda)|^2 d|E(\lambda)f|^2 \leq C^2.$$

Since this is true for every a and b, we have (b). Thus (c) implies (b).

Now (b) implies (a). For if $a' < a < b < b'$ we have

$$|\int_{a'}^{b'} \phi(\lambda)dE(\lambda)f - \int_a^b \phi(\lambda)dE(\lambda)f|^2$$

$$= \int_{a'}^{a} |\phi|^2 d|E(\lambda)f|^2 + \int_b^{b'} |\phi|^2 d|E(\lambda)f|^2$$

by Lemma 3 of the preceding section. This is easily seen to yield that (b) implies (a).

Finally the continuity of the inner product insures that (a) implies (c). These three results yield the equivalence of the three statements.

THEOREM II. If $\phi(\lambda)$ is real and continuous for $-\infty < \lambda < \infty$, $H = \int_{-\infty}^{\infty} \phi(\lambda)dE(\lambda)$ is a self-adjoint transformation, whose domain consists of all f 's such that $\int_{-\infty}^{\infty} |\phi|^2 d|E(\lambda)f|^2 < \infty$ and $|Hf|^2 = \int_{-\infty}^{\infty} |\phi|^2 d|E(\lambda)f|^2$.

PROOF: Hf exists if and only if $\int_{-\infty}^{\infty} |\phi|^2 d|E(\lambda)f|^2 < \infty$

since (a) and (b) in Lemma 2 are equivalent. Lemma 3 of the preceding section implies $|Hf|^2 = \int_{-\infty}^{\infty} |\phi|^2 d|E(\lambda)f|^2$.

We next show that H is symmetric. (Cf. Def. 1 of §3 of Chapter IV) If $f \in \mathfrak{H}$, 3 of Definition 1 of the preceding section implies that there is an a and b such that $|f-(E(b)-E(a))f| < \epsilon$. Lemma 4 of the preceding section implies that $(E(b)-E(a))f$ is in the domain of H. Thus the domain of H is dense. Lemma 3 of the preceding section and the continuity of the inner product imply that $(Hf,g) = (f,Hg)$ for every pair f and g of the domain of H.

Thus $H \subset H^*$. Now let g be in the domain of H^*, and $H^*g = g^*$. Then for every f,

$$(f,(E(b)-E(a))g^*) = ((E(b)-E(a))f,g^*) = ((E(b)-E(a))f,H^*g)$$
$$= (H(E(b)-E(a))f,g) = \int_a^b (\phi)d(E(\lambda)f,g)$$

by Lemma 1 of this section and Lemma 3 of the preceding section. Thus

$$F(f) = \int_{-\infty}^{\infty} \phi(\lambda)d(E(\lambda)f,g) = \lim \int_a^b \phi(\lambda)d(E(\lambda)f,g)$$
$$= \lim (f,(E(b)-E(a))g^*) = (f,g^*)$$

is a bounded linear functional and hence $\int_{-\infty}^{\infty} |\phi|^2 d|E(\lambda)g|^2 < \infty$ by Lemma 2 above. Hence if g is in the domain of H^* it is in the domain of H. This and $H \subset H^*$ imply $H = H^*$.

COROLLARY. If $H = \int_a^b \phi(\lambda)dE(\lambda)$, where $\phi(\lambda)$ is real and continuous for $a < \lambda \leq b$, then H is a self-adjoint transformation defined for all f 's such that $\int_a^b |\phi|^2 d|E(\lambda)f|^2 < \infty$ and $|Hf|^2 = \int_a^b |\phi|^2 d|E(\lambda)f|^2$.

The proof of the corollary is similar to the proof of the Theorem.

LEMMA 3. Let $H = \oint \phi(\lambda)dE(\lambda)$ be an improper integral. Let $\mathfrak{D}_0$ be a subset of the domain of H defined as follows. If $H = \int_{-\infty}^{\infty} \phi(\lambda)dE(\lambda)$, then $\mathfrak{D}_0$ consists of all g 's for which there exists an a_g and a b_g such that $g = (E(b)-E(a))g$. If $H = \int_a^b \phi(\lambda)dE(\lambda)$ for a $\phi(\lambda)$ which is continuous for $a < \lambda \leq b$, then $\mathfrak{D}_0$ consists of all g 's for which there is an $a'_g > a$ such that $(1-E(a'))g = g$.

Let H_0 be the contraction of H with domain $\mathfrak{D}_0$. Then $[H_0] = H$.

Since by the preceding theorem, $H = H^*$, H is closed. Hence $[H_0] \subset H$. On the other hand if $\{f,Hf\}$ is in the graph of H, then given $\epsilon > 0$, we can find an a and a b such that

$$|Hf-H(E(b)-E(a))f| = \int_b^\infty |\phi|^2 d|E(\lambda)f|^2+\int_{-\infty}^a |\phi|^2 d|E(\lambda)f|^2 < \epsilon^2/2$$

and

$$|f-(E(b)-E(a))f|^2 < \epsilon^2/2$$

by Lemma 1 of this section and the preceding Theorem. Now $g = (E(b)-E(a))f$ is in the domain of H_0. Thus the above inequalities imply

$$|\{f,Hf\}-\{g,H_0g\}|^2 < \epsilon^2.$$

Since ϵ is arbitrary, it follows that f is in the domain of $[H_0]$. Hence $H \subset [H_0]$. The inclusions established in this paragraph show that $H = [H_0]$.

The proof is similar in the case of an improper integral of the first kind.

LEMMA 4. Let $H_1 = \int_{-\infty}^\infty \phi_1(\lambda)dE(\lambda)$, $H_2 = \int_{-\infty}^\infty \phi_2(\lambda)dE(\lambda)$ for ϕ_1 and ϕ_2 continuous. Then $H_1 \cdot H_2$ is a transformation with domain, those f's for which $\int_{-\infty}^\infty |\phi_2|^2 d|E(\lambda)f|^2 < \infty$ and $\int_{-\infty}^\infty |\phi_1\phi_2|^2 d|E(\lambda)f|^2 < \infty$. For the contraction given in the preceding Lemma, we have $H_1(H_2)_0 = (H_1)_0 \cdot (H_2)_0 = (H_3)_0$. Finally $[H_1 \cdot H_2] = H_3$. Similar results hold in the case of an improper integral of the first kind.

If f is in the domain of $H_1 \cdot H_2$, then both H_2f and $H_1(H_2f)$ exist. "H_2f exists" is equivalent to $\int_{-\infty}^\infty |\phi_2|^2 d|E(\lambda)f|^2 < \infty$ by Lemma 2 above. On the other hand, when H_2f and H_1H_2f exist,

$$\begin{aligned} H_1(H_2f) &= \lim \int_a^b \phi_1 dE(\lambda)(\int_{-\infty}^\infty \phi_2 dE(\lambda)f) \\ &= \lim \int_a^b \phi_1 dE(\lambda)(E(b)-E(a))\int_{-\infty}^\infty \phi_2 dE(\lambda)f \\ &= \lim (\int_a^b \phi_1 dE(\lambda))(\int_a^b \phi_2(\lambda)dE(\lambda))f \\ &= \lim \int_a^b \phi_1\phi_2 dE(\lambda)f = \int_{-\infty}^\infty \phi_1\phi_2 dE(\lambda)f \end{aligned}$$

by Lemma 4 and 5 of the preceding section and Lemma 1 of this section. Thus f is in the domain of H_3 and we must have $\int_{-\infty}^{\infty} |\phi_1\phi_2|^2 d|E(\lambda)f|^2 < \infty$. On the other hand, if the f 's satisfy both conditions, a similar argument shows that $H_3 f = H_1\cdot(H_2 f)$ and thus f is in the domain of $H_1\cdot(H_2 f)$. We may sum up by stating that the domain of $H_1\cdot H_2$ is the intersection of the domain of H_2 and H_3.

We note concerning the $\mathfrak{D}_0$ of the preceding Lemma, that if $f \in \mathfrak{D}_0$, then by Lemma 1 above $Hf = \int_{-\infty}^{\infty}\phi(\lambda)dE(\lambda)(E(b)-E(a)f) = (E(b)-E(a))\int_{-\infty}^{\infty}\phi(\lambda)dE(\lambda)f = (E(b)-E(a))Hf$. Thus $f \in \mathfrak{D}_0$ implies $Hf \in \mathfrak{D}_0$. It follows then that $H_1(H_2)_0$ is precisely $(H_1)_0(H_2)_0$. Furthermore Lemma 5 of the preceding section and Lemma 1 above yield that if $f \in \mathfrak{D}_0$, then $(H_1)_0\cdot(H_2)_0 f = (H_3)_0 f$. Thus $(H_1)_0\cdot(H_2)_0 = (H_3)_0$.

The first paragraph of this proof shows that $H_1\cdot H_2 \subset H_3$. Thus $[H_1\cdot H_2] \subset H_3$. On the other hand $[H_1\cdot H_2] \supset [(H_1)_0\cdot(H_2)_0] = [(H_3)_0] = H_3$ by the preceding lemma. Hence $[H_1\cdot H_2] = H_3$.

One notes that if ϕ_2 is bounded the condition $\int_{-\infty}^{\infty} |\phi_2|^2 d|E(\lambda)f|^2 < \infty$ is always satisfied. Under such circumstances, one has $H_2\cdot H_1 \subset H_1\cdot H_2 = H_3$.

LEMMA 5. If a continuous $\phi(x)$ is > 0 for $a < x \leqq b$ and $E(a) = 0$, $E(b) = 1$, then $H = \int_a^b \phi(\lambda)dE(\lambda)$ has an inverse $K = \int_a^b \phi(\lambda)^{-1}dE(\lambda)$.

PROOF: In Lemma 4, let $H_1 = H$, $H_2 = K$, $H_3 = \int_a^b dE(\lambda) = E(b)-E(a) = 1$. The remarks immediately preceding the present Lemma show that $HK \subset KH = 1$. Now Lemma 6 of the preceding section shows that H^{-1} exists and $KH = 1$ shows that $H^{-1} \subset K$. $HK \subset 1$ shows that K has an inverse since $Kf = \theta$ implies $1\cdot f = HKf = H\theta = \theta$. Since the range of H^{-1} is $\mathfrak{H}$ however, H^{-1} has no proper extension and thus $H^{-1} = K$.

§4

DEFINITION 1. If T is bounded and H is self-adjoint, and $TH \subset HT$ then T is said to commute with H.

LEMMA 1. If T commutes with H then

(a) T^* commutes with H.

(b) if H is bounded, T commutes with H, H^2 etc.,

and any linear combination of these.

(c) if H has an inverse, T commutes with H^{-1}.

PROOF OF (a): Theorem V of Chapter IV, §2 and its corollary tell us that $HT^* = (TH)^* \supset (HT)^* \supset T^*H$.

PROOF OF (b): Since $TH = HT$ when H is bounded, this is obvious.

PROOF OF (c): $TH \subset HT$ is equivalent to the statement that for every pair $\{f,Hf\}$ in the graph of H, $\{Tf,THf\}$ is in the graph of H. This condition is symmetric for H and H^{-1} and thus if T commutes with H, T commutes with H^{-1}.

LEMMA 2. If T commutes with $\int_a^b \lambda dE(\lambda)$, where $E(a) = 0$, $E(b) = 1$, T commutes with $\int_a^b \phi(\lambda)dE(\lambda)$ where $\phi(\lambda)$ is continuous for $a \leqq x \leqq b$. Also T commutes with $E(\lambda)$ for $a \leqq \lambda \leqq b$.

PROOF: Let $p(x)$ be any polynomial $a_n x^n + a_{n-1}x^{n-1} + \dots + a_0$, and let $p(H) = a_n H^n + a_{n-1}H^{n-1} + \dots + a_0$. Lemma 5 of §2 above and $1 = E(b)-E(a) = \int_a^b dE(\lambda)$ imply $p(H) = \int_a^b p(\lambda)dE(\lambda)$. Lemma 1(b) above shows that T commutes with $p(H)$. If $\phi(x)$ is continuous on the interval $a \leqq x \leqq b$, we can find a sequence of polynomials $p_n(x)$ such that $p_n(x) \longrightarrow \phi(x)$ uniformly in the interval $a \leqq x \leqq b$. Now $\int_a^b \phi(\lambda)dE(\lambda) - p_n(H) = \int_a^b (\phi(\lambda)-p_n(\lambda))dE(\lambda)$. Hence Lemma 3 of §2 above shows that $p_n(H)f \longrightarrow \int_a^b \phi(\lambda)dE(\lambda)f$ for every $f \in \mathfrak{H}$. Thus $T\int_a^b \phi(\lambda)dE(\lambda)f = \lim Tp_n(H)f =$ $\lim p_n(H)Tf = \int_a^b \phi(\lambda)dE(\lambda)Tf$ for every f in $\mathfrak{H}$. Hence T commutes with $\int_a^b \phi(\lambda)dE(\lambda)$.

We next show the last statement of the lemma. Since T obviously commutes with $0 = E(a)$ and $1 = E(b)$, we need only consider the $E(\mu)$ with $a < \mu < b$. For $\epsilon > 0$, let $\phi(x,\mu,\epsilon) = 1$ for $x \leqq \mu$, $\phi(x,\mu,\epsilon) = 1-(x-\mu)/\epsilon$ for $\mu < x \leqq \mu+\epsilon$, $\phi(x,\mu,\epsilon) = 0$ for $x > \mu+\epsilon$. Now $\phi(x,\mu,\epsilon)$ is continuous and hence $H(\mu,\epsilon) = \int_a^b \phi(\lambda,\mu,\epsilon)dE(\lambda)$ commutes with T. Since $\int_a^\mu \phi(\lambda,\mu,\epsilon)dE(\lambda) = E(\mu)-E(a) = E(\mu)$ and $\int_{\mu+\epsilon}^b \phi(\lambda,\mu,\epsilon)dE(\lambda) = 0$ we have $H(\mu,\epsilon) - E(\mu) = \int_\mu^{\mu+\epsilon} \phi(\lambda,\mu,\epsilon)dE(\lambda)$. Then $|(H(\mu,\epsilon)f-E(\mu)f)|^2 = |\int_\mu^{\mu+\epsilon} \phi(\lambda,\mu,\epsilon)dE(\lambda)f|^2 = \int_\mu^{\mu+\epsilon} |\phi|^2 d|E(\lambda)f|^2 \leqq \int_\mu^{\mu+\epsilon} d|E(\lambda)f|^2 = |E(\mu+\epsilon)f|^2 - |E(\mu)f^2$. Since $E(\mu+\epsilon) \longrightarrow E(\mu)$ as $\epsilon \longrightarrow 0$, we must have $H(\mu,\epsilon)f \longrightarrow E(\mu)f$ for every f. Since T commutes with every $H(\mu,\epsilon)$, it is readily seen as in the preceding paragraph to commute with the limit $E(\mu)$.

DEFINITION 2. A c.a.d.d. operator A will be said to be normal if $A^*A = AA^*$.

LEMMA 3. Let A and B be normal and let $A^*A = B^*B$. Let W be such that $A = WB$ as in Theorem VII of §4, Chapter IV. Then W commutes with A^*A. Also the initial set of W is also the final set of W.

PROOF: $WA^*A = WAA^* = WBB^* = AB^* = AA^*W = A^*AW$ when one uses the results of Theorem VII of §4, Chapter IV.

Now $A^{*\prime}$, which is defined in Theorem VIII of §4, Chapter IV has precisely the same zeros as A^*. For $A^{*\prime}$ is a contraction of A^* and hence $\mathfrak{N}_{A^{*\prime}} \subset \mathfrak{N}_{A^*}$. On the other hand, if $A^*f = \theta$, AA^*f exists and hence f is in the domain AA^* which is of course the domain of $A^{*\prime}$. Thus $f \in \mathfrak{N}_{A^*}$ implies $f \in \mathfrak{N}_{A^{*\prime}}$. This and our previous inclusion imply $\mathfrak{N}_{A^*} = \mathfrak{N}_{A^*}$.

Now under our hypotheses $AA^* = A^*A = B^*B = BB^*$. But $AA^*f = \theta$ is equivalent to $A^{*\prime}f = \theta$. The latter obviously implies the former. If however we have $AA^*f = \theta$, then f is in the domain of $A^{*\prime}$ and $0 = (AA^*f,f) = (A^*f,A^*f) = (A^{*\prime}f,A^{*\prime}f)$. Thus $\mathfrak{N}_{A^*} = \mathfrak{N}_{A^{*\prime}} = \mathfrak{N}_{AA^*} = \mathfrak{N}_{BB^*} = \mathfrak{N}_{B^{*\prime}} = \mathfrak{N}_{B^*}$. Since by Theorem VI of Chapter IV, §2, $[\mathfrak{R}_A] = \mathfrak{N}_{A^*}$, we have $[\mathfrak{R}_A] = [\mathfrak{R}_B]$. But these sets are respectively the final and initial sets of W.

LEMMA 4. If A and B are self-adjoint and $A = WB$ where W is partially isometric with initial set $[\mathfrak{R}_B]$ and final set $[\mathfrak{R}_A]$ and W commutes with B, then $W = W^*$.

PROOF: By the corollary to Theorem V of §2, Chapter IV, $A^* = B^*W^*$. Since A and B are self-adjoint, this becomes $A = BW^*$. Lemma 1 above shows that W^* commutes with B. Thus $W^*B \subset BW^* = A = WB$. Since W^*B and WB have the same domain, we have $W^*B = WB$. This means that $W^*g = Wg$ for g in the range of B. Since W^* and W are continuous, we may extend the equation $W^*g = Wg$ to the closure of the range of B.

Now $A = WB$, $A^*A = B^*W^*WB = B^*EB = B^*B$, where E is the projection on $[\mathfrak{R}_B]$ the initial set of W (Cf. Lemma 1 of §3,

Chapter VI). Since A and B are self-adjoint they are normal. Thus the initial set of W is also the final set of W by Lemma 3 above. Lemmas 1 and 3 of §3, Chapter VI show that this set is also the initial set of W^*. Thus W and W^* are zero on the orthogonal complement of the initial set $[\mathfrak{R}_B]$ of W.

Thus $W = W^*$ on $[\mathfrak{R}_B]$ and on $[\mathfrak{R}_B]^{\perp}$. Since W and W^* are additive, we must have $W = W^*$.

CHAPTER VIII

BOUNDED-SELF-ADJOINT AND UNITARY TRANSFORMATIONS

A self-adjoint operator H is said to have an integral representation if

$$H = \int_{-\infty}^{\infty} \lambda dE(\lambda)$$

for a resolution of the identity $E(\lambda)$. (Cf. Chapter VII, §3). A unitary $\mathfrak{U}$ is said to have an integral representation, if $\mathfrak{U} = \int_0^{2\pi} e^{i\mu} dE(\mu)$ for a finite resolution of the identity. (Cf. Chapter VII, §2, Def. 1).

It is a fundamental result in the theory of linear transformations in Hilbert space, that every self-adjoint transformation has an integral representation. Every unitary transformation has likewise. However both of these are normal and these results can even be generalized to the statement that every normal operator has an integral representation.

These general results will be established in Chapter IX. In the present Chapter, we will obtain the integral representation for bounded self-adjoint transformations and for unitary transformations.

§1

In this section, we will suppose that H is self-adjoint and bounded with $|||H||| = C$. If $p(x) = a_n x^n + a_{n-1} x^{n-1} + \ldots + a_0$ is a polynomial, we let $p(H) = a_n H^n + a_{n-1} H^{n-1} + \ldots + a_0$. We consider only the case, when the a's are real, and thus $p(H)$ is self-adjoint. (Cf. Theorem V of §2, Chapter IV. Also Chapter IV, §3, Lemma 4).

LEMMA 1. If for $-C \leqq x \leqq C$, $p(x) \geqq 0$ then $p(H)$ is definite.

Let $\mathfrak{M}$ denote any finite dimensional manifold and E the projection on $\mathfrak{M}$. Let $H' = EHE$. By Theorem V of Chapter IV, §2, $H'^* = (EHE)^* = E^*H^*E^* = EHE$. Hence H' is self-adjoint.

Furthermore $|H'f| = |EHEf| \leq |HEf| \leq C\cdot|Ef| \leq C\cdot|f|$. Thus H' has a bound $\leq C$.

Let $p'(H') = a_nH'^n+a_{n-1}H'^{n-1}+ \dots +a_0E$. Lemma 5 of §1, Chapter VII shows that if f is $\in \mathfrak{M}$, and $p(x) \geq 0$ for $-C \leq x \leq C$, then

$$0 \leq (p'(H')f,f).$$

If f is an arbitrary element of $\mathfrak{H}$ let $\mathfrak{M}$ be the manifold determined by f, Hf, H^2f, ... , $H^{(n)}f$. $\mathfrak{M}$ is finite dimensional. Furthermore $f = Ef$, $Hf = EHf = EHEf = H'f$, $H^2f = EH^2f = EH\cdot Hf = EH\cdot E\cdot Hf = H'\cdot H'f$. A similar argument will show $H^kf = H'^kf$ for $2 \leq k \leq n$. Thus $p(H)f = p'(H')f$. Thus for every f, $(p(H)f,f) = (p'(H')f,f) \geq 0$. This proves the result.

LEMMA 2. If for $-C \leq x \leq C$, $|p(x)| \leq \epsilon$, then $|||p(H)||| \leq \epsilon$.

PROOF: Under these circumstances, Lemma 1 states that $p(H)+\epsilon$ is definite and $\epsilon-p(H)$ is definite. Thus $0 \leq ((p(H)+\epsilon)f,f)$ or $-\epsilon(f,f) \leq (p(H)f,f)$. Hence C_- for $p(H)$ is $\geq -\epsilon$. (Cf. Chapter IV, §3, Definition 4). Similarly $C_+ \leq \epsilon$. Lemma 10 of Chapter IV, §3 now implies that $|||p(H)||| \leq \epsilon$.

LEMMA 3. If $P(x)$ is a continuous function on the interval $-C \leq x \leq C$, then there exists a unique bounded operator, $P(H)$ such that for every $\epsilon > 0$ and every polynomial $p(x)$ such that $|P(x)-p(x)| < \epsilon$ for $-C \leq x \leq C$ then $|||P(H)-p(H)||| < \epsilon$.

PROOF: The existence of one such $P(H)$ can be established as follows. Let $\epsilon_1, \epsilon_2, \dots, \epsilon_n > 0$ be a sequence of positive numbers such that $\epsilon_n \longrightarrow 0$ as $n \longrightarrow \infty$. Since $P(x)$ is continuous, we can find for each ϵ_n a polynomial $p_n(x)$ such that $|P(x)-p_n(x)| < \epsilon_n$ for $-C \leq x \leq C$. This implies that for $-C \leq x \leq C$, $|p_n(x)-p_m(x)| \leq |p_n(x)-P(x)|+|P(x)-p_m(x)| < \epsilon_n+\epsilon_m$. Lemma 2 above shows that $|||p_n(H)-p_m(H)||| \leq \epsilon_n+\epsilon_m$. Hence for every $f \in \mathfrak{H}$, $|p_n(H)f-p_m(H)f| \leq (\epsilon_n+\epsilon_m)\cdot|f|$ and the sequence $p_n(H)f$ is convergent to some f^*. Let $P(H)f = f^*$. The sequence $p_n(H)$ is easily shown to be uniformly bounded by Lemma 1 above. Hence $P(H)$ is bounded.

We prove that $P(H)$ has the property given in the Lemma. For suppose $|P(x)-p(x)| < \epsilon$ for $-C \leq x \leq C$. Since P and p are continuous, we can find a $\delta > 0$ such that $|(P(x)-p(x))| < \epsilon-\delta$. Let n_0 be such that for $n > n_0$, $\epsilon_n < \delta$. Then if $n > n_0$ we have $|p_n(x)-P(x)| < \epsilon_n < \delta$ for $-C \leq x \leq C$. (See preceding paragraph). Since we also have $|P(x)-p(x)| \leq \epsilon-\delta$, it follows that $|p_n(x)-p(x)| < \epsilon$. Thus for every f, $|(p_n(H)-p(H))f| < \epsilon|f|$. This holds for every $n > n_0$. Letting $n \longrightarrow \infty$ we get $|P(H)f-p(H)f| < \epsilon|f|$. This implies $|||P(H)-p(H)||| < \epsilon$.

$P(H)$ is unique. For suppose we have two distinct operators P_1 and P_2 with the given properties. Since given $\epsilon > 0$, we can find a $p(x)$ with $|P(x)-p(x)| < \epsilon$ we must have $|||P_i(H)-p(H)||| < \epsilon$ for $i = 1, 2$. Hence $|||P_1(H)-P_2(H)||| < 2\epsilon$. Since ϵ is arbitrary, this implies $|||P_1-P_2||| = 0$.

LEMMA 4. If $P(x)$ and $P(H)$ are as in Lemma 3 above then (a): If $P(x) = p_0(x)$ is a polynomial, then $P(H) = p_0(H)$. (b): If $P(x) \geq 0$ for $-C \leq x \leq C$, then $P(H)$ is definite. (c): If $|P(x)| \leq k$ for $-C \leq x \leq C$, then $|||P(H)||| \leq k$.

(a) is a consequence of the uniqueness of $P(H)$.

We next show (b). Let ϵ_α and $p_\alpha(x)$ be as in the first paragraph of the proof of Lemma 3 above. Let $q_\alpha(x) = p_\alpha(x)+\epsilon_\alpha$. Then $q_\alpha(x) = p_\alpha(x)+\epsilon_\alpha \geq P(x) \geq 0$. Hence $q_\alpha(H)$ is definite by Lemma 1 above. For every f, $\lim q_\alpha(H)f = \lim p_\alpha(H)f + \lim \epsilon_\alpha f = P(H)f$. Hence $P(H)$ is the limit of definite operators and can be shown to be symmetric, self-adjoint and definite.

(c) is proved in a manner analogous to the proof of Lemma 2 above.

LEMMA 5. Let $P(x)$, $Q(x)$ and $R(x)$ be continuous functions and let $P(H)$, $Q(H)$ and $R(H)$ denote the corresponding operators in Lemma 3. Then (a): If $R(x) = P(x)+Q(x)$ for $-C \leq x \leq C$, then $R(H) = P(H)+Q(H)$. (b): If $R(x) = P(x)\cdot Q(x)$ then $R(H) = P(H)\cdot Q(H)$.

PROOF OF (a): Given $\epsilon > 0$, we can find polynomials $p(x)$ and $q(x)$ such that $|P(x)-p(x)| < \epsilon/2$, $|Q(x)-q(x)| < \epsilon/2$ for $-C \leq x \leq C$. Then $|R(x)-(p(x)+q(x))| = |P(x)-p(x)+Q(x)-q(x)| < \epsilon$

on the same interval. Lemma 3 now implies that $|||R(H)-(p(H)+q(H))||| < \epsilon$, $|||P(H)-p(H)||| < \epsilon/2$ and $|||Q(H)-q(H)||| < \epsilon/2$. These inequalities imply $|||R(H)-(P(H)+Q(H))||| < 2\epsilon$. Since ϵ is arbitrary we must have $|||R(H)-(P(H)+Q(H))||| = 0$, $R(H) = P(H)+Q(H)$.

PROOF OF((b): Let k be such that $|P(x)|+1 < k$, $|Q(x)| < k$ for $-C \leq x \leq 1$. Let ϵ be such that $0 < \epsilon \leq 1$. We can find polynomials $p(x)$ and $q(x)$ such that $k|P(x)-p(x)| < \epsilon$ and $k|Q(x)-q(x)| < \epsilon$ for $-C \leq x \leq C$. Since $k > 1$, we have $|P(x)-p(x)| < \epsilon/k < \epsilon$ which implies $|p(x)| < |P(x)|+\epsilon \leq |P(x)| +1 < k$, for the given x -interval. Lemma 4 above then implies that $|||p(H)||| < k$. Lemma 4 and the other inequalities imply $|||Q(H)||| < k$, $k|||P(H)-p(H)||| < \epsilon$ and $k|||Q(H)-q(H)||| < \epsilon$. Now for $-C \leq x \leq C$, $|R(x)-p(x)q(x)| = |P(x)Q(x)-p(x)q(x)| = |P(x)\cdot Q(x)-p(x)Q(x)+p(x)Q(x)-p(x)q(x)| \leq |Q(x)|\cdot|P(x)-p(x)|+|p(x)|\cdot |Q(x)-q(x)| < k|P(x)-p(x)|+k|Q(x)-q(x)| < 2\epsilon$. Thus Lemma 3 implies that $|||R(H)-p(H)q(H)||| < 2\epsilon$. We also have $|||P(H)Q(H)-p(H)q(H)||| = |||P(H)Q(H)-p(H)Q(H)+p(H)Q(H)-p(H)q(H)||| \leq |||(P(H)-p(H))Q(H)||| + |||p(H)(Q(H)-q(H))||| \leq |||P(H)-p(H)|||\cdot|||Q(H)||| + |||p(H)|||\cdot|||Q(H)-q(H)||| < k|||P(H)-p(H)||| +k|||Q(H)-q(H)||| < 2\epsilon$. This and the preceding result imply $|||R(H)-P(H)Q(H)||| < 4\epsilon$. Since ϵ is arbitrary, we must have $R(H) = P(H)Q(H)$.

If A is c.a.d.d., we let $\mathfrak{N}_A$ denote the zeros of A. If $\mathfrak{R}_A$ is the range of A and A is self-adjoint then $[\mathfrak{R}_A] = \mathfrak{N}_A^{\perp}$. (Cf. Lemma 8 of §3, Chapter IV). For definite operators, we also have the result.

LEMMA 6. If A and B are self-adjoint bounded and definite, then $\mathfrak{N}_{A+B} \subset \mathfrak{N}_A$ and consequently $[\mathfrak{R}_A] \subset [\mathfrak{R}_{A+B}]$.

PROOF: Let f be such that $(A+B)f = \theta$. Then $0=((A+B)f,f) = (Af,f)+(Bf,f)$. Since $(Af,f) \geq 0$ and $(Bf,f) \geq 0$, this implies $(Af,f) = 0$.

For every f and g, we also have the result $|(Af,g)|^2 \leq (Af,f)(Ag,g)$. Inasmuch as A is definite, we have for every real λ and every complex z with $|z| = 1$,

$$(A(f+\lambda zg),f+\lambda zg) \geq 0.$$

Expanding and using the symmetry of A, we get

$$(Af,f)+2\lambda R(\overline{z}(Af,g))+\lambda^2(Ag,g) \geq 0.$$

We may take z so that $R(\overline{z}(Af,g)) = |(Af,g)|$. Hence

$$(Af,f)+2\lambda|(Af,g)|+\lambda^2(Ag,g) \geq 0.$$

If $(Ag,g) = 0$, this implies $|(Af,g)| = 0$ and our desired inequality holds in this case. If $(Ag,g) \neq 0$, let $\lambda = -|(Af,g)|/(Ag,g)$ and the above inequality becomes

$$(Af,f)-\frac{|(Af,g)|^2}{(Ag,g)} \geq 0.$$

Hence $(Af,f)(Ag,g) \geq |(Af,g)|^2$.

From the preceding two paragraphs, we see that if $(A+B)f=\theta$, $(Af,f) = 0$, and hence $(Af,g) = 0$ for every $g \in \mathfrak{H}$. Hence $Af = \theta$. Hence $f \in \mathfrak{N}_{A+B}$ implies $f \in \mathfrak{N}_A$ or $\mathfrak{N}_{A+B} \subset \mathfrak{N}_A$. The orthogonality relations mentioned before the Lemma, yield $[\mathfrak{R}_{A+B}] \supset [\mathfrak{R}_A]$.

§2

We now consider two functions $P_\mu(x) = \max(x-\mu,0)$, $Q_\mu(x) = \max(-(x-\mu),0)$. $P_\mu(x)$ and $Q_\mu(x)$ are continuous, non-negative, $Q_\mu(x)P_\mu(x) = P_\mu(x)Q_\mu(x) = 0$ and $P_\mu(x)-Q_\mu(x) = x-\mu$. Applying Lemmas 3, 4 and 5 of the preceding section, we obtain:

LEMMA 1. Let $P_\mu(x)$, $Q_\mu(x)$ be as above and let $P_\mu(H)$, $Q_\mu(H)$ be the corresponding operators as in Lemma 3. Then $P_\mu(H)$ and $Q_\mu(H)$ are definite, $Q_\mu(H)P_\mu(H) = P_\mu(H)Q_\mu(H) = 0$ and $P_\mu(H)-Q_\mu(H) = H-\mu$.

LEMMA 2. Let $\mathfrak{N}(\mu) = \mathfrak{N}_{P_\mu(H)}$ and $E(\mu)$ be the projection on $\mathfrak{N}(\mu)$. Then (a): If $\mu_1 > \mu_2$, $E(\mu_1) \geq E(\mu_2)$. (b): $\lim_{\epsilon\to 0}E(\mu+\epsilon^2) = E(\mu)$. (c): $E(C) = 1$, $E(-C-\epsilon) = 0$, for $\epsilon > 0$.

PROOF: If $\mu_1 > \mu_2$, $P_{\mu_2}(x)-P_{\mu_1}(x) \geq 0$ and $P_{\mu_1}(x) \geq 0$. Lemmas 4 and 5 of the preceding section imply that $P_{\mu_2}(H)-P_{\mu_1}(H)$ and $P_{\mu_1}(H)$ are definite. Lemma 5 of the preceding section implies that $P_{\mu_2}(H) = (P_{\mu_2}(H)-P_{\mu_1}(H))+P_{\mu_1}(H)$. Lemma 6 of that section now implies that

$$\mathfrak{N}_{P_{\mu_1}} \supset \mathfrak{N}_{P_{\mu_2}} \quad \text{or} \quad E(\mu_1) \geq E(\mu_2).$$

We next prove (b). We first show that for every $f \in \mathfrak{H}$, $\lim_{\epsilon \to 0} P_{\mu+\epsilon^2}(H)f = P_\mu(H)f$. One notes that $0 \leqq P_\mu(x) - P_{\mu+\epsilon^2}(x) \leqq \epsilon^2$. Lemma 4 of the preceding section implies that $|||P_\mu(H) - P_{\mu+\epsilon^2}(H)||| \leqq \epsilon^2$. From this, we may easily infer that $\lim_{\epsilon \to 0} P_{\mu+\epsilon^2}(H)f = P_\mu(H)f$ for every f.

We also need that $\lim_{\epsilon \to 0} E(\mu+\epsilon^2)$ exists and is the projection on $\Pi_\epsilon \mathfrak{N}(\mu+\epsilon^2)$. If $\epsilon_1 > \epsilon_2 > \ldots$ is a sequence with $\lim_{n\to\infty} \epsilon_n = 0$, then Lemma 10 of §1, Chapter VI implies that $\lim_{n\to\infty} E(\mu+\epsilon_n^2)$ exists and is a projection E with range $\Pi_n \mathfrak{N}(\mu+\epsilon_n^2)$. If ϵ is such that $\epsilon_{n+1}^2 \leqq \epsilon^2 \leqq \epsilon_n^2$, then from the above, we know that $E(\mu+\epsilon_{n+1}^2) \leqq E(\mu+\epsilon^2) \leqq E(\mu+\epsilon_n^2)$. From Lemma 6 of Chapter VI, §1, we know that if $\mu_1 > \mu_2$, then $E(\mu_1) - E(\mu_2)$ is a projection and our preceding result permits us to state that $E(\mu+\epsilon_{n+1}^2) - E \leqq E(\mu+\epsilon^2) - E \leqq E(\mu+\epsilon_n^2) - E$. Lemma 8 of §1, Chapter VI now implies that for every f, $|E(\mu+\epsilon_{n+1}^2)f - Ef| \leqq |E(\mu+\epsilon^2)f - Ef| \leqq |E(\mu+\epsilon_n^2)f - Ef|$. It follows that $\lim_{\epsilon \to 0} E(\mu+\epsilon^2)$ exists and is E. Since $\mathfrak{N}(\mu+\epsilon_{n+1}^2) \subset \mathfrak{N}(\mu+\epsilon^2) \subset \mathfrak{N}(\mu+\epsilon_n^2)$, we also have $\Pi_\epsilon \mathfrak{N}(\mu+\epsilon^2) = \Pi_n \mathfrak{N}(\mu+\epsilon_n^2)$.

Thus $\lim_{\epsilon \to 0} E(\mu+\epsilon^2) = E(\mu)$ if $\Pi_\epsilon \mathfrak{N}(\mu+\epsilon^2) = \mathfrak{N}(\mu)$. Since $\mathfrak{N}(\mu) \subset \mathfrak{N}(\mu+\epsilon^2)$, we must have $\Pi_\epsilon \mathfrak{N}(\mu+\epsilon^2) \supset \mathfrak{N}(\mu)$. But if $f \in \Pi_\epsilon \mathfrak{N}(\mu+\epsilon^2)$ then $f \in \mathfrak{N}(\mu+\epsilon^2)$ and $P_{\mu+\epsilon^2}(H)f = \theta$. Letting $\epsilon \to 0$, we obtain by a result given above that $P_\mu(H)f = \theta$ or $f \in \mathfrak{N}(\mu)$. Thus $f \in \Pi_\epsilon \mathfrak{N}(\mu+\epsilon^2)$ implies $f \in \mathfrak{N}(\mu)$. This and $\mathfrak{N}(\mu) \subset \Pi_\epsilon \mathfrak{N}(\mu+\epsilon^2)$ imply $\mathfrak{N}(\mu) = \Pi_\epsilon \mathfrak{N}(\mu+\epsilon^2)$ and $E(\mu) = \lim_{\epsilon \to 0} E(\mu+\epsilon^2)$.

$P_C(x) = 0$ for $-C \leqq x \leqq C$. Hence $P_C(H)f = \theta$ for every $f \in \mathfrak{H}$ and $E(C) = 1$. If $\epsilon > 0$, then $P_{-C-\epsilon}(x) - \epsilon \geqq 0$ for $-C \leqq x \leqq C$ and hence $((P_{-C-\epsilon}(H) - \epsilon)f, f) \geqq 0$, by Lemma 4 of the preceding section. Thus for every f, $(P_{-C-\epsilon}(H)f, f) \geqq \epsilon(f,f)$ and $P_{-C-\epsilon}(H)f = \theta$ implies $f = \theta$. Thus $\mathfrak{N}(-C-\epsilon) = \{\theta\}$ and $E(-C-\epsilon) = 0$.

Definition 1 of §2, Chapter VII now informs us that the $E(\mu)$'s form a resolution of the identity.

COROLLARY. The $E(\mu)$'s of Lemma 2 form a resolution of the identity.

LEMMA 3. The $E(\mu)$'s of Lemma 2 above commute with H. Consequently if $\mu_2 < \mu_1$, $H(E(\mu_1) - E(\mu_2)) = (E(\mu_1) - E(\mu_2))H(E(\mu_1) - E(\mu_2))$.

PROOF: Lemma 5 of the preceding section implies that $HP_\mu(H)$ and $P_\mu(H)H$ both correspond to $x \cdot P_\mu(x) = P_\mu(x) \cdot x$. Hence $HP_\mu(H) = P_\mu(H)H$ and if $P_\mu(H)f = \theta$, $\theta = HP_\mu(H)f = P_\mu(H)Hf$. Thus $f \in \mathfrak{N}(\mu)$ implies $Hf \in \mathfrak{N}(\mu)$. It follows that $HE(\mu) = E(\mu)HE(\mu)$. Taking adjoints yields $E(\mu)H = E(\mu)HE(\mu) = HE(\mu)$.

Thus $H(E(\mu_1)-E(\mu_2)) = HE(\mu_1)-HE(\mu_2) = E(\mu_1)H-E(\mu_2)H = (E(\mu_1)-E(\mu_2))H$. Multiplying this equation on the right by $E(\mu_1)-E(\mu_2)$ will give the desired result.

It is convenient to denote the interval $\mu_2 < x \leqq \mu_1$ by I and then let $E(I) = E(\mu_1)-E(\mu_2)$.

LEMMA 4. If $f \in \mathfrak{N}(\mu)$, $(Hf,f) \leqq \mu(f,f)$. If $f \in \mathfrak{N}(\mu)^\perp$, $(Hf,f) \geqq \mu(f,f)$.

PROOF: If $f \in \mathfrak{N}(\mu)$, $P_\mu(H)f = \theta$. Since $(P_\mu(H)-Q_\mu(H))f = Hf-\mu f$, we must have $\mu f-Hf = Q_\mu(H)f$. But Lemma 1 above also tells us that $Q_\mu(H)f$ is definite. Hence $(\mu f-Hf,f) = (Q_\mu(H)f,f) \geqq 0$. Thus $\mu(f,f) \geqq (Hf,f)$.

If $f \in \mathfrak{N}(\mu)^\perp$, $f \in [\mathfrak{R}_{P_\mu(H)}]$ by Lemma 8 of §3, Chapter IV. Lemma 1 above states that $Q_\mu(H)P_\mu(H) = 0$. Thus $\mathfrak{N}_{Q_\mu(H)} \supset [\mathfrak{R}_{P_\mu(H)}]$ and $Q_\mu(H)f = \theta$. Since $(P_\mu(H)-Q_\mu(H))f = Hf-\mu f$, we have $P_\mu(H)f = Hf-\mu f$. Lemma 1 above states that $P_\mu(H)$ is definite and thus $(Hf-\mu f,f) = (P_\mu(H)f,f) \geqq 0$. Hence $(Hf,f) \geqq \mu(f,f)$.

LEMMA 5. For every $f \in \mathfrak{H}$ and $\epsilon > 0$, we have $(Hf,f) = \int_{-C-\epsilon}^{C} \lambda d(E(\lambda)f,f)$.

PROOF: Divide the interval $-C-\epsilon \leqq \lambda \leqq C$ into subintervals by the points $\mu_0 = -C-\epsilon, < \mu_1 < \dots < \mu_{n-1} < \mu_n = C_n$. Let I_α denote the interval $\mu_{\alpha-1} \leqq x \leqq \mu_\alpha$ and $E(I_\alpha) = E(\mu_\alpha)-E(\mu_{\alpha-1})$. Lemma 6 of §1, Chapter VI, tells us that the range of $E(I_\alpha)$ is $\mathfrak{N}(\mu_\alpha)\mathfrak{N}(\mu_{\alpha-1})^\perp$. Lemma 4 above now implies that

$$\mu_{\alpha-1}(E(I_\alpha)f,f) = \mu_{\alpha-1}(E(I_\alpha)f,E(I_\alpha)f) \leqq (HE(I_\alpha)f,E(I_\alpha)f)$$
$$\leqq \mu_\alpha(E(I_\alpha)f,E(I_\alpha)f) = \mu_\alpha(E(I_\alpha)f,f).$$

We also have $\Sigma_{\alpha=1}^{n} E(I_\alpha) = E(C)-E(-C-\epsilon) = 1-0$, by Lemma 2 above. Lemma 3 above then shows that

$$(Hf,f) = (H(\Sigma_{\alpha=1}^{n} E(I_\alpha))f,f) = \Sigma_{\alpha=1}^{n}(HE(I_\alpha)f,f)$$

$$= \Sigma^n_{\alpha=1}(E(I_\alpha)HE(I_\alpha)f,f) = \Sigma^n_{\alpha=1}(HE(I_\alpha)f,E(I_\alpha)f).$$

The previous inequality on $(HE(I_\alpha)f,E(I_\alpha)f)$ now yields that

$$\Sigma^n_{\alpha=1}\mu_{\alpha-1}(E(I_\alpha)f,f) \leqq (Hf,f) \leqq \Sigma^n_{\alpha=1}\mu_\alpha(E(I_\alpha)f,f).$$

Lemma 8 of §1, Chapter VI, shows that $(E(\mu)f,f) = |E(\mu)f|^2$ is a monotonically increasing function of μ. Thus the right and left hand expressions in the preceding inequalities can be made to approach the desired integral simultaneously and our Lemma is proven.

§3

If we let $H_1 = \int^C_{-C-\epsilon}\lambda dE(\lambda)$ by using Definition 3 of §2, Chapter VII, then Lemma 3 of §2, Chapter VII implies that $(H_1f,f) = \int^C_{-C-\epsilon}\lambda d(E(\lambda)f,f)$. Lemma 5 of the preceding section states that $(Hf,f) = \int^C_{-C-\epsilon}\lambda d(E(\lambda)f,f)$ and thus for every $f \in \mathfrak{H}$, $(H_1f,f) = (Hf,f)$ or $((H_1-H)f,f) = 0$. H_1-H is symmetric with domain $\mathfrak{H}$ and hence self-adjoint. From Definition 4 of §3, Chapter IV we see that $C_+ = 0$, $C_- = 0$, for H_1-H and Lemma 10 which follows this definition yields $|||H_1-H||| = 0$ or $H = H_1$. We have proved:

THEOREM I. If H is self-adjoint with bound C and $\epsilon > 0$, then there exists a finite resolution of the identity $E(\lambda)$ such that

$$H = \int^C_{-C-\epsilon}\lambda dE(\lambda).$$

We can refine our result somewhat by using the considerations of §2, Chapter VII, which follow Lemma 4. If C_+ for H is $< C$, we take a δ so small that $C_++\delta < C$. In the discussion referred to, we let $\lambda_1 = C_++\delta, \lambda_2 = C$. We then obtain that $\int^C_{-C-\epsilon}\lambda dE(\lambda) = \int^{C_++\delta}_{-C-\epsilon}\lambda dE(\lambda)$ for every $\delta > 0$ and $E(C)-E(C_++\delta)=0$. Since $E(C) = 1$, we have $E(C_++\delta) = 1$. Now $|\int^{C_++\delta}_{C_+}\lambda dE(\lambda)f|^2 = \int^{C_++\delta}_{C_+}\lambda^2 d|E(\lambda)f|^2 \leqq k^2|(E(C_++\delta)-E(C_+))f|^2$ where $k = \max(|C_+|, |C_++\delta|)$, when we use Lemma 3 of §2, Chapter VII. Since $E(C_++\delta)f \longrightarrow E(C_+)f$ when we let $\delta \longrightarrow 0$, $\delta > 0$, by Lemma 2 of the preceding section, we must have that the $\lim_{\delta\to 0,\delta>0} \int^{C_++\delta}_{C_+}\lambda dE(\lambda)f$ exists and $= \theta$. This and our preceding result imply $\int^C_{-C-\epsilon}\lambda dE(\lambda) = \int^{C_+}_{-C-\epsilon}\lambda dE(\lambda)$. We also have $E(C_+) = E(C_++0) = 1$. Of course if $C_+ = C$ these results are still true.

If we consider the lower limit of integration, we have to consider that $\lim_{\delta\to 0, \delta>0}\int_{C_- -\delta}^{C_-}\lambda dE(\lambda) = C_-(E(C_-)-E(C_- -0))$. But the discussion which follows Lemma 4 of §2, Chapter VII shows also that $E(C_- -\delta)-E(-C-\epsilon) = 0$ for $C_- -\delta > -C-\epsilon$ and $\delta > 0$. Since $E(-C-\epsilon) = 0$, $E(C_- -\delta) = 0$ for $\delta > 0$ and hence $E(C_- -0) = 0$. Thus $\lim_{\delta\to 0,\delta>0}\int_{C_- -\delta}^{C_-}\lambda dE(\lambda) = C_-E(C_-)$. A proof similar to that of the preceding paragraph will now show that $\int_{-C-\epsilon}^{C_+}\lambda dE(\lambda) = \int_{C_-}^{C_+}\lambda dE(\lambda)+C_-E(C_-)$.

COROLLARY. If C_+ and C_- refer to H as in Definition 4 of §3, Chapter IV, then

$$H = \int_{C_-}^{C_+}\lambda dE(\lambda)+C_-E(C_-)$$

and $E(C_+) = 1$, $E(C_- -0) = 0$.

§4

In this section, we will deal with Unitary Transformations. (Cf. Definition 1 of §2, Chapter VI). From Lemma 1 of §2, Chapter VI, we know that $\mathfrak{U}\mathfrak{U}^* = \mathfrak{U}^*\mathfrak{U} = 1$. Let $A = \frac{1}{2}(\mathfrak{U}+\mathfrak{U}^*)$, $B = \frac{1}{2}i(\mathfrak{U}^*-\mathfrak{U})$. Then $iB = \frac{1}{2}(\mathfrak{U}-\mathfrak{U}^*)$ and $\mathfrak{U} = A+iB$. Theorem V of §2 of Chapter IV implies that A and B are self-adjoint and $A-iB = \mathfrak{U}^*$. We also have that $AB = \frac{1}{4}i(\mathfrak{U}+\mathfrak{U}^*)(\mathfrak{U}^*-\mathfrak{U}) = \frac{1}{4}i(\mathfrak{U}^{*2}-\mathfrak{U}^2) = \frac{1}{4}i(\mathfrak{U}^*-\mathfrak{U})(\mathfrak{U}+\mathfrak{U}^*) = BA$. Thus $1 = (A+iB)(A-iB) = A^2 + B^2$.

We shall obtain an integral representation for $\mathfrak{U}$ by using the integral representation for A given by Theorem I of the preceding section. It will however be convenient to assume that $\mathfrak{U}f = f$ or $\mathfrak{U}f = -f$ implies $f = \theta$. We shall see that this assumption can be made without an essential loss of generality.

Let $\mathfrak{M}_1$ denote the set of f 's for which $\mathfrak{U}f=f$, and $\mathfrak{M}_{-1}$ denote the set of f 's for which $\mathfrak{U}f = -f$. $\mathfrak{M}_1$ and $\mathfrak{M}_{-1}$ are mutually orthogonal since if $f \in \mathfrak{M}_1$ and $g \in \mathfrak{M}_2$, $(f,g) = (\mathfrak{U}f,\mathfrak{U}g) = (f,-g) = -(f,g)$ or $2(f,g) = 0$. If E_1 and E_{-1} are the projections on $\mathfrak{M}_1$ and $\mathfrak{M}_{-1}$ then $E_1E_{-1} = 0$ and E_1+E_{-1} is the projection with range $\mathfrak{U}(\mathfrak{M}_1 \cup \mathfrak{M}_{-1})$ by Lemma 5 of §1, Chapter VII. Let $F = 1-(E_1+E_{-1})$.

By the definition of $\mathfrak{M}_1$, we have $\mathfrak{U}E_1 = E_1$. Since $\mathfrak{U}^*= \mathfrak{U}^{-1}$, $f \in \mathfrak{M}_1$ implies $\mathfrak{U}^*f = f$ and hence $\mathfrak{U}^*E_1= E_1$. Taking adjoints we obtain $E_1\mathfrak{U} = E_1= \mathfrak{U}E_1$. Similarly $E_{-1}\mathfrak{U} = -E_{-1}= \mathfrak{U}E_{-1}$. These results also imply that $F\mathfrak{U} = \mathfrak{U}F$. Multiplying on the left by F yields $F\mathfrak{U} = F\mathfrak{U}F$.

Thus

$$\mathfrak{U} = \mathfrak{U}(E_1+E_{-1}+F) = \mathfrak{U}E_1+\mathfrak{U}E_{-1}+\mathfrak{U}F = E_1-E_{-1}+F\mathfrak{U}F.$$

If we contract $F\mathfrak{U}F$ to the range of F, we see that for every f and g in this range, $(F\mathfrak{U}Ff,F\mathfrak{U}Fg) = (\mathfrak{U}Ff,\mathfrak{U}Fg) = (\mathfrak{U}f,\mathfrak{U}g) = (f,g)$. Thus $F\mathfrak{U}F$ is isometric on the range of F.

Furthermore the range of $F\mathfrak{U}F$ on the range of F is also the range of F. Otherwise there would be a $g \neq \theta$ in the range of F such that for every $f \in \mathfrak{H}$, $0 = (g,F\mathfrak{U}F(Ff)) = (g,F\mathfrak{U}Ff) = (g,F\mathfrak{U}f) = (Fg,\mathfrak{U}f) = (g,\mathfrak{U}f)$. Thus g is orthogonal to the range of $\mathfrak{U}$ and since $\mathfrak{U}$ is unitary, $g = \theta$. This is a contradiction. Thus we have proved that the range of $F\mathfrak{U}F$ on the range of F is the range of F.

Definition 1 of §2, Chapter VI, is now satisfied by $F\mathfrak{U}F$ regarded as a transformation on the range of F i.e., in this sense $F\mathfrak{U}F$ is unitary. For g in the range of F, $F\mathfrak{U}Fg = g$ is equivalent to $\mathfrak{U}Fg = g$ or $\mathfrak{U}g = g$. Thus $g \in \mathfrak{M}_1$. Since g is also in the range of F, $g = \theta$. Hence $F\mathfrak{U}Fg = g$, g in the range of F, implies $g = \theta$. Similarly $F\mathfrak{U}Fg = -g$, g in the range of F, implies $g = \theta$.

Thus we have proved:

LEMMA 1. Let $\mathfrak{U}$, $\mathfrak{M}_1$, $\mathfrak{M}_{-1}$, E_1, E_{-1}, and F be defined as above. Then $\mathfrak{U}E_1 = E_1 = E_1\mathfrak{U}$, $E_{-1}\mathfrak{U} = -E_{-1} = \mathfrak{U}E_{-1}$, $F\mathfrak{U}F=\mathfrak{U}F=F\mathfrak{U}$. The ranges of E_1, E_{-1} and F are orthogonal. We also have $\mathfrak{U} = E_1-E_{-1}+F\mathfrak{U}F$. Let $\mathfrak{U}_1$ denote $F\mathfrak{U}F$ considered as a transformation on the range of F. Then $\mathfrak{U}_1$ is unitary and $\mathfrak{U}_1 g = g$ or $\mathfrak{U}_1 g = -g$ implies $g = \theta$.

In applying Definition 1 of §2, Chapter VI, we have assumed that the range of F is infinite dimensional i.e., a Hilbert space. (Cf. the corollary to Theorem II of §2, Chapter III). The cases in which the range of F is finite can be easily taken care of by elementary methods. We consider then only the case in which the range of F is infinite dimensional.

For a time we will consider only those $\mathfrak{U}$'s for which $\mathfrak{U}f = f$ or $\mathfrak{U}f = -f$ implies $f = \theta$. Let A and B be as in the first paragraph of this section. Since $A = \frac{1}{2}(\mathfrak{U}+\mathfrak{U}^*)$, $\|A\| \leq \frac{1}{2}(\|\mathfrak{U}\| + \|\mathfrak{U}^*\|)$. Thus Theorem I of the preceding section shows that there is a resolution of the identity $E(\lambda)$ such that for $\epsilon > 0$, we have

$$A = \int_{-1-\epsilon}^{1} \lambda dE(\lambda).$$

As a proof of the corollary of that theorem shows, we may write this

$$A = \int_{-1}^{1} \lambda dE(\lambda) - E(-1).$$

But if f is in the range of $E(-1)$, $Af = -f$. Since $A^2 + B^2 = 1$, we have $(Af,Af) = (f,f) = ((A^2+B^2)f,f) = (A^2f,f) + (B^2f,f) = (Af,Af)+(Bf,Bf)$. Hence $|Bf|^2 = (Bf,Bf) = 0$ and $Bf = \theta$. Thus $\mathfrak{U}f = (A+iB)f = Af+iBf = Af = -f$. But this implies $f = \theta$. Thus if f is in the range of $E(-1)$, $f = \theta$ and $E(-1) = 0$.

The expression for A also shows that if f is in the range of $E(1)-E(+1-0)$, $Af = f$. An argument similar to that of the preceding paragraph now implies $f = \theta$. Thus $E(1)-E(+1-0) = 0$. But $E(1) = 1$ and hence $E(1-0) = 1$. We have shown:

LEMMA 2. If $\mathfrak{U}$ is unitary and such that $\mathfrak{U}f = f$ or $\mathfrak{U}f = -f$ implies $f = \theta$, then if $A = \frac{1}{2}(\mathfrak{U}+\mathfrak{U}^*)$ there is a resolution of the identity $E(\lambda)$ such that

$$A = \int_{-1}^{1} \lambda dE(\lambda),$$

and $E(-1) = 0$, $E(1-0) = 1$.

From Lemma 5 of §2, Chapter VII, we have that $B^2 = 1-A^2 = \int_{-1}^{1}(1-\lambda^2)dE(\lambda)$. Let $C = \int_{-1}^{1}(1-\lambda^2)^{1/2}dE(\lambda)$; Lemma 5 of §2, Chapter VII implies $C^2 = \int_{-1}^{1}(1-\lambda^2)dE(\lambda) = B^2$. C and B are self-adjoint and thus we may apply Theorem I of §4, Chapter VI to obtain that there is a partially isometric W such that $B = WC$.

For $C = \int_{-1}^{1}(1-\lambda^2)^{1/2}dE(\lambda)$, the discussion which follows Lemma 6 of §2, Chapter VII shows that the zeros of C form the range of $1-E(1-0)+E(-1) = 0$. Thus $\mathfrak{N}_C = \{\theta\}$ and C^{-1} exists. A discussion similar to the proof of Lemma 5 of §3, Chapter VII will show that $C^{-1} = \int_{-1}^{1}(1-\lambda^2)^{-1/2}dE(\lambda)$. The corollary to Theorem II of §3, Chapter VII now shows that the domain of C^{-1} is dense.

Lemma 5 of §2, Chapter VII shows that $AC = \int_{-1}^{1}\lambda(1-\lambda^2)^{1/2}dE(\lambda) = CA$. Thus A commutes with C. (Cf. Definition 1 of §4, Chapter VII). Lemma 1(c) of §4, Chapter VII implies that A commutes with C^{-1}. In the first paragraph of this section it was shown that $AB = BA$. Thus $ABC^{-1} = BAC^{-1} \subset BC^{-1}A$. Since $WC = B$, $BC^{-1}f = Wf$ for f in the domain of C^{-1}. Thus for f in the domain of C^{-1}, $AWf = WAf$. Since W and A are continuous and the domain of C^{-1} is dense, we must have $AW = WA$.

Since W and A commute, Lemma 2 of §4, Chapter VII implies that W commutes with $E(\lambda)$ for $-1 \leqq \lambda \leqq 1$. Since W commutes with $E(\lambda)$ for $-1 \leqq \lambda \leqq 1$, it is readily seen to commute with any $\Sigma_{\Pi'}\phi\Delta E(\lambda)$ as defined in Definition 2 of §2, Chapter VII. W will then commute with any limit of these, $\int_{-1}^{1}\phi(\lambda)dE(\lambda)$. In particular W will commute with $C = \int_{-1}^{1}(1-\lambda^2)^{1/2}dE(\lambda)$. W will also commute with $B = WC$.

Thus we have proved:

LEMMA 3. Let A, B, C and $E(\lambda)$ be as above. There exists a partially isometric W with initial set $[\mathfrak{R}_C]$ and final set $[\mathfrak{R}_B]$ such that $B = WC$. W commutes with A, B, C and $E(\lambda)$. $[\mathfrak{R}_C] = \mathfrak{H}$.

With regard to the last statement, we recall that we have shown that $\mathfrak{R}_C$ which is the domain of C^{-1} is dense.

When we apply Lemma 4 of §4, Chapter VII we obtain:

COROLLARY 1. $W = W^*$.

COROLLARY 2. If T is bounded and commutes with $E(\lambda)$ for a resolution of the identity, T commutes with $\int_a^b\phi(\lambda)dE(\lambda)$ for every $\phi(\lambda)$ for which this integral exists.

LEMMA 4. Let W be partially isometric, $W = W^*$ and let $E = W^*W$. Then $F_1 = \frac{1}{2}(E+W)$ and $F_2 = \frac{1}{2}(E-W)$ are projections such that $F_1+F_2 = E$, $F_1-F_2 = W$.

F_1 and F_2 are self-adjoint, since E and W are. Since $W = W^*$, $E = W^*W = W^2$ and $EW = W^3 = WE = W$ since E is the initial set of W. (Cf. Lemma 1 of §3, Chapter VI). Thus $F^2 = \frac{1}{4}(E+W)(E+W) = \frac{1}{4}(E^2+WE+EW+W^2) = \frac{1}{4}(2E+2W) = \frac{1}{2}(E+W) = F$. Similarly $F_2^2 = F_2$. Lemma 2 of §1, Chapter VI now implies that F_1 and F_2 are projections.

For W in Lemma 3 above, E is the projection on $[\mathfrak{R}_C] = \mathfrak{H}$ and thus $E = 1$. Since W and 1 commute with $E(\lambda)$ for $-1 \leqq \lambda \leqq 1$, F_1 and F_2 also commute with $E(\lambda)$ for $-1 \leqq \lambda \leqq 1$. Now

$$\begin{aligned} \mathfrak{U} = A+iB &= A+iWC = (F_1+F_2)A+i(F_1-F_2)C = F_1(A+iC)+F_2(A-iC) \\ &= F_1\int_{-1}^{1}(\lambda+i(1-\lambda^2)^{1/2})dE(\lambda)+F_2\int_{-1}^{1}(\lambda-i(1-\lambda^2)^{1/2})dE(\lambda) \\ &= \int_{-1}^{1}(\lambda+i(1-\lambda^2)^{1/2})dF_1E(\lambda)+\int_{-1}^{1}(\lambda-i(1-\lambda^2)^{1/2})dF_2E(\lambda). \end{aligned}$$

For ϕ between 0 and π let us define $F(\phi) = F_1-F_1E(\cos\phi -0)$. Since F_1 and the $E(\lambda)$ commute $F_1E(\cos\phi-0)$ is a projection. (Cf. Lemma 4 of §1, Chapter VI). Since $F_1E(\cos\phi-0) \leqq F_1$, Lemma 6 of the same section shows that $F(\phi)$ is a projection. One easily verifies that if $\phi_1 < \phi_2$, $F(\phi_1) \leqq F(\phi_2)$ because of the analogous property of the $E(\lambda)$. One also has $\lim_{\epsilon\to 0}F(\phi+\epsilon^2) = F(\phi)$, $F(0) = F_1-F_1E(1-0) = F_1-F_1 = 0$, $F(\pi) = F_1-F_1E(-1-0) = F_1$.

It remains to show that $\int_0^{\pi}e^{i\phi}dF(\phi) = \int_{-1}^{1}(\lambda+i(1-\lambda^2)^{1/2})dF_1E(\lambda)$. There is at most a denumerable set of points at which $E(\lambda) \neq E(\lambda-0)$. If we avoid these points in forming partitions and use $\lambda = \cos\phi$, the partial sums for each integral are the same and thus the limits are equal or $\int_0^{\pi}e^{i\phi}dF(\phi) = \int_{-1}^{1}(\lambda+i(1-\lambda^2)^{1/2})dF_1E(\lambda)$.

For ϕ between π and 2π, we let $F(\phi) = F_1+F_2E(\cos\phi)$. The discussion of the preceding paragraphs can be extended and we see that $F(\phi)$ is a projection, $F(\phi_1) \leqq F(\phi_2)$ if $\phi_1 < \phi_2$, $\lim_{\epsilon\to 0}F(\phi+\epsilon^2) = F(\phi)$, $F(\pi) = F_1$, ($F(\pi)$ is defined in two ways) $F(2\pi) = F_1+F_2 = 1$ and

$$\begin{aligned} \int_{-1}^{1}(\lambda-i(1-\lambda^2)^{1/2})dF_2E(\lambda) &= \int_{-1}^{1}(\lambda-i(1-\lambda^2)^{1/2})d(F_1+F_2E(\lambda)) \\ &= \int_{\pi}^{2\pi}e^{i\phi}dF(\phi). \end{aligned}$$

If we combine our results, we obtain,

$$\mathfrak{U} = \int_0^{2\pi}e^{i\phi}dF(\phi).$$

We may sum up as follows:

LEMMA 5. Let $\mathfrak{U}$ be a unitary transformation such that $\mathfrak{U}f = f$ or $\mathfrak{U}f = -f$ implies $f = \theta$. Then there exists a resolution of the identity $F(\phi)$, with $F(0) = 0$, $F(2\pi) = 1$, such that

$$\mathfrak{U} = \int_0^{2\pi}e^{i\phi}dF(\phi).$$

To remove the restrictions on $\mathfrak{U}$, we recall the situation described in Lemma 1 above. We apply Lemma 5 to $\mathfrak{U}_1 = F\mathfrak{U}F$ considered on the range of F and obtain a family of projections $F(\phi)$ of the range of F. Now either Definition 1 or Lemma 2 of §1, Chapter VI, can be used to show that $F(\phi)F$ is a projection of $\mathfrak{H}$. The orthogonality of F, E_1 and E_{-1} insures that $E_{-1}+F(\phi)F$ and $E_{-1}+E_1+F(\phi)F$ are projections. We define $G(\phi) = F(\phi)F$ for $0 \leqq \phi < \pi$, $G(\phi) = E_{-1}+F(\phi)F$ for $\pi \leqq \phi < 2\pi$ and $G(2\pi) = E_1+E_{-1}+F = 1$. One can easily verify that the $G(\phi)$ form a resolution of the identity with $G(0) = 0$, $G(2\pi) = 1$. Furthermore we have that

$$\int_0^{2\pi} e^{i\phi} dG(\phi) = \left(\int_0^{2\pi} e^{i\phi} dF(\phi)\right)F+E_1+E_{-1} = F\mathfrak{U}F+E_1-E_{-1} = \mathfrak{U}.$$

Thus we have established:

THEOREM II. If $\mathfrak{U}$ is unitary, there exists a resolution of the identity $G(\phi)$ with $G(0) = 0$, $G(2\pi) = 1$, such that

$$\mathfrak{U} = \int_0^{2\pi} e^{i\phi} dG(\phi).$$

CHAPTER IX

CANONICAL RESOLUTION AND INTEGRAL REPRESENTATIONS

In this Chapter, we obtain the canonical resolution of a c.a. d.d. operator T and the integral representations of self-adjoint and normal operators.

The discussion of Chapters VIII, IX and X is essentially based on the two papers of J. von Neumann to which reference is made at the end of Chapter I. In the present Chapter, however the use of $(1+H^2)^{-1}$ to obtain the integral representation of an unbounded self-adjoint operator was suggested by the Riesz-Lorch paper also listed in Chapter I. The canonical resolution of a normal operator is used to obtain the integral representation by K. Kodaira.*

§1

In this section, we obtain the canonical resolution of a c.a. d.d. operator T. (Cf. Theorem I of this section), For a c.a. d.d. operator T, Theorem VII of §4, Chapter IV tells us that $A = (1+T^*T)^{-1}$ is a bounded definite self-adjoint operator with a bound ≤ 1. The Corollary to Theorem I of §3, Chapter VIII shows that there exists a resolution of the identity $E(\lambda)$ with $E(0-0)$, $E(1) = 1$, such that

$$A = \int_0^1 \lambda dE(\lambda) + 0 \cdot E(0) = \int_0^1 \lambda dE(\lambda).$$

Lemma 6 of §2, Chapter VII shows that the zeros of A are $1-E(1)+E(0) = E(0)$. Since A^{-1} exists, $\mathfrak{N}_A = \{\theta\}$ and hence $E(0) = 0$.

Lemma 5 of §3, Chapter VII implies that $A^{-1} = \int_0^1 (1/\lambda) dE(\lambda)$. We make the change of variable $\mu = 1/\lambda$, $F(\mu) = 1-E(1/\mu-0)$ for $1 \leq \mu < \infty$. Since $E(0+0) = E(0) = 0$ and $\lim_{\lambda\to 0, \lambda>0} E(\lambda-0) = \lim_{\lambda\to 0, \lambda>0} E(\lambda) = \lim_{\mu\to 0} F(\mu) = 1$, $F(1) = 1-E(1-0)$. As in the discussion of the proof of Lemma 5 of §4, Chapter VIII, one can show that $F(\mu)$ is a resolution of the identity and that for $1 \leq b < \infty$,

$$\int_{1/b}^1 (1/\lambda) dE(\lambda) = \int_1^b \mu dF(\mu) + F(1).$$

---

* Proc. Imp. Acad., Todyo 15, pp. 207-210, (1939).

When we let $b \longrightarrow \infty$, we see that corresponding improper integrals are equal and thus

$$A^{-1} = \int_0^1 (1/\lambda) dE(\lambda) = \int_1^\infty \mu dF(\mu) + F(1).$$

Thus

$$1+T^*T = \int_1^\infty \mu dF(\mu) + F(1)$$

or

$$T^*T = \int_1^\infty (\mu-1) dF(\mu).$$

We make the further change of variable $\mu'^2 = \mu - 1$, $F_1(\mu') = F(\mu'^2+1)$ for $0 \leqq \mu' < \infty$, $F_1(\mu') = 0$, for $\mu' < 0$. We then obtain

$$T^*T = \int_0^\infty \mu'^2 dF_1(\mu).$$

Let $B = \int_0^\infty \mu' dF_1(\mu')$. $\int_0^\infty \mu'^4 d|E(\lambda)f|^2 < \infty$ implies $\int_0^\infty \mu'^2 d|E(\lambda)f|^2 < \infty$. Hence Lemma 4 of §3, Chapter VII shows that $B^2 = \int_0^\infty \mu'^2 dF_1(\mu') = T^*T$. B is self-adjoint by Theorem II of §3, Chapter VII. Furthermore $B^*B = B^2 = T^*T$. Thus Theorem I of §4, Chapter VI states that there is a partially isometric W with initial set $[\mathfrak{R}_B]$ and final set $[\mathfrak{R}_T]$ and such that $T = WB$, $T^* = B^*W^* = BW^*$, $B = W^*T = T^*W$.

LEMMA 1. Let T be c.a.d.d. Then there exists a resolution of the identity $F_1(\mu)$ with $F_1(0) = 0$ and such that if $B = \int_0^\infty \mu dF_1(\mu)$, then $B^2 = T^*T$, $T = WB$, $T^* = BW^*$, $B = W^*T = TW$. Furthermore if $E(\lambda) = 1-F_1((1/\lambda-1)^{1/2}-0)$, then $(1+T^*T)^{-1} = \int_0^1 \lambda dE(\lambda)$.

The equation $T = WB$ is called the canonical resolution. For completeness, we must still consider the corresponding results for T^*. Lemma 1 above when applied to T^* shows that there is a resolution of the identity $F_2(\mu)$ and $C = \int_0^\infty \mu dF_2(\mu)$ such that $C^2 = TT^*$.

LEMMA 2. If D is a bounded transformation, which commutes with $B^2 = T^*T$, then D commutes with $F_1(\lambda)$ and and B. A similar result holds for C.

PROOF: If D commutes with B^2, it also commutes with B^2+1 and $(B^2+1)^{-1}$. (Cf. Lemma 1 of §4, Chapter VII). Lemma 2 of §4, Chapter VII now shows that D commutes with $E(\lambda)$ for $0 \leqq \lambda \leqq 1$.

By Lemma 1 above, we see that this implies that D commutes with $F_1(\mu)$. By Corollary 2 to Lemma 3 of Chapter VIII, §4, we obtain that D commutes with $\int_0^a \mu dF_1(\mu)$ for every a. Thus if f is in the domain of D, we have $\int_0^a \mu dF_1(\mu)Df = D(\int_0^a \mu dF_1(\mu))f$. D is continuous and hence the limit as $a \longrightarrow \infty$ of the right hand side exists. Hence Df is in the domain of B whenever f is, and $DB \subset BD$.

LEMMA 3. Let T, B, C, W, $F_1(\lambda)$ and $F_2(\lambda)$ be as above. Then $WF_1(\lambda)W^* = F_2(\lambda)-F_2(0)$, $W^*F_2(\lambda)W = F_1(\lambda)-F_1(0)$, $WBW^* = C$, $B = W^*CW$.

Let $E_1 = W^*W$, $E_2 = WW^*$. These are projections on, respectively, the initial and final sets of W, i.e., $[\mathfrak{R}_B]$ and $[\mathfrak{R}_T]$. Now if we use $[\mathfrak{N}_A]^\perp = [\mathfrak{R}_A]$, (Theorem VI of §2, Chapter IV) and $B = B^*$, $C = C^*$, $\mathfrak{N}_T = \mathfrak{N}_B$, $\mathfrak{N}_C = \mathfrak{N}_{T^*}$, we obtain $[\mathfrak{R}_B] = \mathfrak{N}_B^\perp$, $[\mathfrak{R}_T] = \mathfrak{N}_T^\perp = \mathfrak{N}_C^\perp$. Thus E_1 is the projection on $\mathfrak{N}_B^\perp$, E_2 on $\mathfrak{N}_C^\perp$. From the expression for B and C given in Lemmas 1 and 2 above, we have that $\mathfrak{N}_B$ has the projection $F_1(0)$ and $\mathfrak{N}_C$ has the projection $F_2(0)$ (Cf. proof of Lemma 6 of §2, Chapter VII). Hence $E_1 = 1-F_1(0)$, $E_2 = 1-F_2(0)$.

Thus E_α commutes with F_α, $\alpha = 1, 2$. From Lemma 3 of §3, Chapter VI we see that $W = WW^*W = E_2W = WE_1$.

Now $WF_1(\lambda)W^*$ is bounded and self-adjoint. Furthermore $WF_1(\lambda)W^* = WE_1F_1(\lambda)W^* = WE_1F_1^2(\lambda)W^* = WF_1(\lambda)E_1F_1(\lambda)W^* = WF_1(\lambda)W^*WF_1(\lambda)W^* = (WF_1(\lambda)W^*)^2$. Thus $F_2'(\lambda) = WF_1(\lambda)W^*$ is a projection. (Cf. Lemma 2 of §1, Chapter VI). Since $E_2F_2'(\lambda) = E_2WF_1(\lambda)W^* = WF_1(\lambda)W^* = F_2'(\lambda)$, we have $F_2'(\lambda) \leqq E_2$. (Cf. Lemma 8 of §1, Chapter VI).

If we define $F_1'(\lambda) = W^*F_2(\lambda)W$, a similar argument will show that $F_1'(\lambda)$ is a projection with $F_1'(\lambda) \leqq E_1$.

We next observe that $C^2 = T^*T = WB\cdot BW^* = WB^2W^*$. Since E_1 is the projection on $[\mathfrak{R}_B]$ and on $\mathfrak{N}_B^\perp$, we have $W^*C^2W = W^*WB^2W^*W = E_1B^2E_1 = B^2$.

Since $C^2 = WB^2W^*$, we have $F_2'(\lambda)C^2 = WF_1(\lambda)W^*WB^2W^* = WF_1(\lambda)E_1B^2W^* = WF_1(\lambda)B^2W^* \subset W(\int_0^\lambda \mu^2 dF_1(\mu))W^* = WB^2F_1(\lambda)W^* = WB^2E_1F_1(\lambda)W^* = WB^2W^*WF_1(\lambda)W^* = C^2F_2'(\lambda)$. (Lemmas 1 and 4 of §3, Chapter VII are used here). Thus $F_2'(\lambda)C^2 \subset C^2F_2'(\lambda)$ and the latter has bound λ^2. Similarly $F_1'(\lambda)B^2 \subset B^2F_1'(\lambda)$ and the latter has bound λ^2.

Lemma 2 above now shows that $F_2'(\lambda)$ commutes with $F_2(\mu)$. We note that $F_2'(\lambda) \leqq E_2 = 1-F_2(0)$. Consider $F_2'(\lambda)(1-F_2(\lambda))$. Lemma 4 of §1, Chapter VI shows that this is a projection. Suppose that an $f \neq \theta$ is in the range of this projection. For a resolution of the identity, we have $\lim_{\mu\to\infty} F_2(\mu) = 1$ and $\lim_{\epsilon\to 0} F_2(\lambda+\epsilon^2) = F_2(\lambda)$. Since $f = F_2'(\lambda)(1-F_2(\lambda))f \neq \theta$, it follows that we can find a Λ and a λ_0 with $\lambda < \lambda_0 < \Lambda$ such that $g = F_2'(\lambda)(F_2(\Lambda)-F_2(\lambda_1))f \neq \theta$. Owing to the commutativity of $F_2'(\lambda)$ and $F_2(\mu)$, we also have $g = (F_2(\Lambda)-F_2(\lambda_0))g = F_2'(\lambda)g$. Hence $|C^2g|^2 = |C^2(F_2(\Lambda)-F_2(\lambda_0))g|^2 = \int_{\lambda_0}^{\Lambda} \lambda^4 d|F_2(\lambda)g|^2 \geqq \lambda_0^4|(F_2(\Lambda)-F_2(\lambda_0))g|^2 = \lambda_0^4|g|^2$. (Cf. Lemma 3 of §2, Chapter VII and Lemma 1 of §3, Chapter VII). Also $|C^2g|^2 = |C^2F_2'(\lambda)g|^2 \leqq \lambda^4|g|^2$. Since $|g| \neq 0$, $\lambda_0 > \lambda \geqq 0$, these statements contradict each other and thus $f = \theta$. Thus we have shown that f in the range of $F_2'(\lambda)(1-F_2(\lambda))$ implies $f = \theta$. It follows that $F_2'(\lambda)(1-F_2(\lambda)) = 0$ or $F_2'(\lambda) \leqq F_2(\lambda)$. (Cf. Lemma 8 of §2, Chapter VI). Since we also have $F_2'(\lambda) \leqq E_2 = 1-F_2(0)$, we have $F_2'(\lambda)(F_2(\lambda)-F_2(0)) = F_2'(\lambda)(F_2(\lambda)-F_2(\lambda)F_2(0)) = F_2'(\lambda)F_2(\lambda)(1-F_2(0))$ $= F_2'(\lambda)(1-F_2(0)) = F_2'(\lambda)$, and $F_2'(\lambda) \leqq F_2(\lambda)-F_2(0)$. (Cf. Definition 1 of §2, Chapter VII and Lemma 8 of §2, Chapter VI).

Now $WF_1(0)W^* = WE_1F_1(0)W^* = W(1-F_1(0))F_1(0)W^* = 0$. Similarly $W^*F_2(0)W = 0$. Thus $F_2'(\lambda) \leqq F_2(\lambda)-F_2(0)$ becomes $F_2(\lambda)-F_2(0) \geqq WF_1(\lambda)W^* = W(F_1(\lambda)-F_1(0))W^*$. Multiplying by W^* on the left and W on the right, we obtain $F_1'(\lambda) = W^*(F_2(\lambda)-F_2(0))W$ $\geqq W^*W(F_1(\lambda)-F_1(0))W^*W = E_1(F_1(\lambda)-F_1(0))E_1 = (1-F_1(0))(F_1(\lambda)-F_1(0))(1-F_1(0)) = F_1(\lambda)-F_1(0)$, or $F_1'(\lambda) \geqq F_1(\lambda)-F_1(0)$. (Cf. Definition 1 of §2, Chapter VII). But a proof analogous to that of the preceding paragraph will show that $F_1'(\lambda) \leqq F_1(\lambda)-F_1(0)$. Thus we have established that $F_1'(\lambda) = F_1(\lambda)-F_1(0)$.

This last result may be written $W^*(F_2(\lambda)-F_2(0))W = F_1(\lambda)-F_1(0)$. Multiplying by W on the right, and on the left and proceeding as above we obtain $F_2(\lambda)-F_2(0) = W(F_1(\lambda)-F_1(0))W^* = F_2'(\lambda) = WF_1(\lambda)W^*$.

If we form partial sums, use this last equation and pass to the limit, we obtain

$$\int_0^\lambda \mu dF_2(\mu) = W(\int_0^\lambda \mu dF_1(\mu))W^*$$

since either side is defined everywhere. Taking limits, we get

$$\int_0^\infty \mu dF_2(\mu) = W(\int_0^\infty \mu dF_1(\mu))W^*$$

or $C = WBW^*$. Multiplying on the left by W^* and on the right by W yields $W^*CW = W^*WBW^*W = E_1BE_1 = B$. Our lemma is now demonstrated.

LEMMA 4. Let T be c.a.d.d. There is at most one B and W such that W is partially isometric with initial set $[\mathfrak{R}_B]$ and final set $[\mathfrak{R}_T]$ and such that B is definite self-adjoint and possesses a resolution of the identity $F_1(\lambda)$ such that $B = \int_0^\infty \mu dF_1(\mu)$ and furthermore such that $T = WB$.

Suppose that the pair, W, $B = \int_0^\infty \mu dF_1(\mu)$ and the pair, W_1, $B_1 = \int_0^\infty \mu dG_1(\mu)$ both satisfy the given conditions. We can suppose that W and B are as in Lemma 1 above. By the corollary to Theorem V of §2, Chapter IV, $T^* = B_1^*W^* = B_1W^*$ and since W^*W is the projection on $[\mathfrak{R}_B]$, we must have $T^*T = B_1^2$. (Cf. Lemma 1 of §3, Chapter VI).

Thus $B_1^2 = T^*T = B^2$ and $\int_0^\infty \mu^2 dG_1(\mu) = \int_0^\infty \mu^2 dF_1(\mu)$ by Lemma 4 of §3, Chapter VII. Now $G_1(\lambda)$ commutes with B_1^2 by Lemma 1 of §3, Chapter VII and thus with B^2. Lemma 2 above, shows that $G_1(\lambda)$ commutes with $F_1(\mu)$ and if we consider the bound of $B_1^2G_1(\lambda)$ we see that precisely the same argument as that used in the proof of Lemma 3 above will show that $G_1(\lambda)-G_1(0) \leqq F_1(\lambda)-F_1(0)$.

We also have in the above that $F_1(\mu)$ commutes with $G_1(\lambda)$ and thus we may proceed to obtain the symmetric result $F_1(\lambda)-F_1(0) \leqq G_1(\lambda)-G_1(0)$. Thus we have shown $F_1(\lambda)-F_1(0) = G_1(\lambda)-G_1(0)$. Since $\mathfrak{N}_{B^2} = \mathfrak{N}_{B_1^2}$, we also have $F_1(0) = G_1(0)$. We may conclude that $F_1(\lambda) = G_1(\lambda)$ and $B = B_1$.

Since $[\mathfrak{R}_B] = [\mathfrak{R}_{B_1}]$, the initial sets of W and W_1 are the same and both W and W_1 are zero on $[\mathfrak{R}_B]^\perp$. (Cf. Definition 1 of §3, Chapter VI). The equation $WB = T = W_1B$ shows that $W = W_1$ on $\mathfrak{R}_B$. Continuity implies $W = W_1^*$ on $[\mathfrak{R}_B]$. We also have $W = W_1 = 0$ on $[\mathfrak{R}_B]^\perp$ and since these transformations are linear we must have $W = W_1$.

COROLLARY. A similar result holds for $T = CW$.

We now complete our discussion:

$C = WBW^*$ implies $CW = WB = T$. The corollary to Theorem V of §2, Chapter IV shows that $T^* = W^*C = BW^*$.

THEOREM I. Let T be c.a.d.d. Then there exists operators W, B, C and resolutions of the identity $F_1(\lambda)$ and $F_2(\lambda)$ such that (a): W is partially isometric with initial set $[\mathfrak{R}_B]$ and final set $[\mathfrak{R}_T]$. (b): B and C are self-adjoint and definite. (c): $T = WB = CW$. (d): $T^* = BW^* = W^*C$. (e): $C = WBW^*$, $B = W^*CW$. (f): $B = \int_0^\infty \lambda dF_1(\lambda)$, $C = \int_0^\infty \lambda dF_2(\lambda)$. (a), (b), (c) and (f) determine W, B and C uniquely.

We shall show in the next section that (b) implies (f). We can then state that (a), (b) and (c) determine W, B and C uniquely.

§2

We now obtain the integral representation for a self-adjoint operator H. H is c.a.d.d. and thus we may apply Theorem I of the preceding section to obtain that there is a definite self-adjoint B with an integral representation, $\int_0^\infty \mu dF_1(\mu)$ a definite, self-adjoint C and a partially isometric W such that $H = WB = CW$. Since $H^*H = H^2 = HH^*$ we see from Lemmas 1 and 2 of the preceding section that $B = C$ and thus $H = WB = BW$.

Thus W commutes with C and Lemma 4 of §4, Chapter VII shows that $W = W^*$. Let $E = W^*W$. Since $W = W^*$, $W = WE = W^3 = EW$. Since W commutes with B, W commutes with B^2 and thus Lemma 2 of the preceding section shows that W commutes with $F_1(\lambda)$. In the proof of the same Lemma 2, it was also shown that $E = 1-F_1(0)$ and consequently E also commutes with $F_1(\lambda)$.

Thus if $F_1 = \frac{1}{2}(E+W)$, $F_2 = \frac{1}{2}(E-W)$ then F_1 and F_2 commute with $F_1(\lambda)$. F_1 and F_2 are projections by Lemma 4 of §4 Chapter VIII. $F_1(\lambda)F_1$ and $F_1(\lambda)F_2$ are projections by Lemma 4 of §1, Chapter VI. We also have $F_1F_2 = \frac{1}{4}(E+W)(E-W) = \frac{1}{4}(E^2+WE-EW-W^2) = \frac{1}{4}(E+W-W-E) = 0$.

We have then:

$$H = BW = \int_0^\infty \mu dF_1(\mu)(F_1 - F_2) = \int_0^\infty \mu dF_1(\mu)F_1 - \int_0^\infty \mu dF_1(\mu)F_2.$$

Let $\lambda = -\mu$, $G(\lambda) = (1 - F_1(-\lambda - 0))F_2$ for $-\infty < \lambda < 0$. Then $\lim_{\lambda \to -\infty} G(\lambda) = \lim_{\mu \to \infty} (1 - F_1(\mu - 0))F_2 = F_2 - \lim_{\mu \to \infty} F_1(\mu)F_2 = F_2 - F_2$ $= 0$ and $\lim_{\epsilon \to 0} G(-\epsilon^2) = \lim (1 - F_1(\epsilon^2 - 0))F_2 = \lim_{\epsilon \to 0} (1 - F_1(\epsilon^2))F_2$ $= (1 - F_1(0 + 0))F_2 = (1 - F_1(0))F_2 = EF_2 = (F_1 + F_2)F_2 = F_2$. As in the proof of Lemma 5 of §4, Chapter VIII, we have that for $b > \epsilon^2 > 0$,

$$-\int_{\epsilon^2}^{b-0} \mu dF_1(\mu)F_2 = \int_{-b}^{-\epsilon^2} \lambda dG_1(\lambda).$$

Letting $\epsilon \to 0$, we have

$$-\int_0^{b-0} \mu dF_1(\mu)F_2 = \lim_{\epsilon \to 0} \int_{-b}^{-\epsilon^2} \lambda dGG(\lambda) = \int_{-b}^{0-0} \lambda dG_1(\lambda).$$

Let $G_1(0) = F_2 + (1 - E) = G_1(0 - 0) + 1 - E$. Since $(1 - E)F_2 = (1 - (F_1 + F_2))F_2 = 0$, $G_1(0)$ is a projection by Lemma 5 of §1, Chapter VI. Also

$$\int_{0-0}^0 \lambda dG_1(\lambda) = 0 \cdot (G_1(0) - G(0 - 0)) = 0.$$

For any integral $\int_{-b}^0 = \int_{-b}^{-\epsilon^2} + \int_{-\epsilon^2}^0 = \lim_{\epsilon \to 0} (\int_{-b}^{-\epsilon^2} + \int_{-\epsilon^2}^0) = \int_{-b}^{0-0} + \int_{0-0}^0$, when these last limits exist. Hence

$$\int_{-b}^0 \lambda dG_1(\lambda) = \int_{-b}^{0-0} \lambda dG_1(\lambda) + \int_{0-0}^0 \lambda dG_1(\lambda) = -\int_0^{b-0} \mu dF_1(\mu)F_2 + 0.$$

Letting $b \to \infty$, we obtain

$$-\int_0^\infty \mu dF_1(\mu)F_2 = \int_{-\infty}^0 \lambda dG_1(\lambda).$$

For $0 < \lambda < \infty$, let $G_1(\lambda) = F_2 + 1 - E + F_1 F(\lambda)$. Our previous orthogonality relations and Lemma 5 of §1, Chapter VI, shows that $G_1(\lambda)$ is a projection. One has also $\lim_{\lambda \to \infty} G_1(\lambda) = F_2 + 1 - E + \lim_{\lambda \to \infty} F_1 F_1(\lambda) = F_2 + 1 - E + F_1 = 1$, since $E = F_2 + F_1$.

By a familiar reasoning, we obtain that

$$\int_0^\infty \lambda dF_1(\lambda)F = \int_0^\infty \lambda dG_1(\lambda).$$

This and our preceding results imply that

$$H = \int_0^\infty \mu dF_1(\mu)F_1 - \int_0^\infty \mu dF_1(\mu)F_2 = \int_{-\infty}^\infty \lambda dG_1(\lambda).$$

Further $G_1(\lambda)$ is a resolution of the identity. For as we have seen above each $G_1(\lambda)$ is a projection and $\lim_{\lambda \to -\infty} G_1(\lambda) = 0$, $\lim_{\lambda \to \infty} G_1(\lambda) = 1$. The other properties of a resolution of the identity are the results of known inclusion relations on the given projections and the corresponding properties for $F_1(\mu)$. We have proved:

THEOREM II. If H is self-adjoint, then there exists a resolution of the identity $G_1(\lambda)$ such that

$$H = \int_{-\infty}^{\infty} \lambda dG_1(\lambda).$$

COROLLARY 1. If $C_- > -\infty$, we have that

$$H = \int_{C_-}^{C_+} \lambda dG_1(\lambda) + C_- G_1(C_-),$$

while if $C_- = -\infty$, we have

$$H = \int_{-\infty}^{C_+} \lambda dG_1(\lambda).$$

This is shown by a discussion similar to the proof of the Corollary to Theorem I of §3, Chapter VIII.

For a definite operator A, we have $C_- \geq 0$. In any case, we may take $0, \infty$ as our limits of integration and $A = \int_0^{\infty} \lambda dG_1(\lambda)$. In Theorem I of §1 above, we now have that (b) implies (f). This demonstrates:

COROLLARY 2. In Theorem I of §1, above (a), (b) and (c) determine W, B and C uniquely.

COROLLARY 3. If T commutes with H, T commutes with $G_1(\lambda)$ for $-\infty < \lambda < +\infty$.

PROOF: If T commutes with H, it commutes with $H^2 = B^2$. It follows from Lemma 2 of the preceding section that T commutes with $F_1(\lambda)$ and with B.

We next show that T commutes with the projection $1-E$ on $\mathfrak{N}_B = \mathfrak{N}_H$. For if f is in $\mathfrak{N}_H$, $Hf = \theta$ and $HTf = THf = T\theta = \theta$ or $Tf \in \mathfrak{N}_H$. Hence $T(1-E) = (1-E)T(1-E)$. By Lemma 1 of §4, Chapter VII, H also commutes with T^* and we obtain $T^*(1-E) = (1-E)T^*(1-E)$. Taking adjoints, we find that $(1-E)T = (1-E)T(1-E) = T(1-E)$. It follows that $ET = TE = ETE$.

In the proof of Theorem II above, we have shown that $W = WE$. Thus if f is in the range of $1-E$, $TWf = TWEf = \theta$ and $WTf = WETf = WTEf = \theta$. Hence for f in the range $1-E$, $TWf = \theta = WTf$.

For f in the domain of B which is also the domain of H, we have $TWBf = THf = HTf = WBTf = WTBf$. Thus if g is in the range of B, $TWg = WTg$. Since W and T are continuous, we may infer that $TWg = WTg$ for $g \in [\mathfrak{R}_B]$, which is also the range of E. This and the result of the preceding paragraph permit us to conclude that $TW = WT$.

If F_1 and F_2 are as in the proof of Theorem II above, we have $F_\alpha T = TF_\alpha$ for $\alpha = 1, 2,$ since $TW = WT$ and $ET = TE$. In the above, we have shown that T commutes with $F_1(\lambda)$. From the definition of $G_1(\lambda)$, we may now infer that T commutes with $G_1(\lambda)$.

COROLLARY 4. The resolution of the identity $G_1(\lambda)$ of Theorem II above is unique.

Let us suppose that H also equals $\int_{-\infty}^{\infty} \lambda dG_2(\lambda)$ for a resolution of the identity $G_2(\lambda)$. We prove that $G_1(\lambda) = G_2(\lambda)$.

For a given λ, $G_2(\lambda)$ commutes with H and thus from Corollary 3 above, it must commute with $G_1(\mu)$ for $-\infty < \mu < \infty$. Thus $G_2(\lambda)(1-G_1(\lambda))$ is a projection by Lemma 4 of §1, Chapter VI. Suppose now that $f \neq \theta$ is in the range of $G_2(\lambda)(1-G_1(\lambda))$. Since $G_1(\lambda+0) = G_1(\lambda)$, $\lim_{\mu\to\infty} G_1(\mu) = 1$, $\lim_{\mu\to-\infty} G_2(\mu) = 0$, we can find three numbers $\lambda_1 < \lambda < \lambda_2 < \lambda_3$ such that $g = (G_2(\lambda)-G_2(\lambda_1))(G_1(\lambda_3)-G_1(\lambda_2))f \neq \theta$.

Now g is easily seen to be in the domain of H. (Cf. Theorem II of §3, Chapter VII). Let a be defined by the equation $a(g,g) = (Hg,g)$. (We recall that $(g,g) \neq 0$). Since $G_1(\lambda)$ and $G_2(\lambda)$ commute we have $g = (G_2(\lambda)-G_2(\lambda_1))g = (G_1(\lambda_3)-G_1(\lambda_2))g$. Lemma 1 of §3, Chapter VII now shows that $(Hg,g) = (H(G_2(\lambda)-G_2(\lambda_1))g,g) = \int_{\lambda_1}^{\lambda}\mu d(G_2(\mu)g,g) \leq \lambda((G_2(\lambda)g,g)-G_2(\lambda_1)g,g) = \lambda((G_2(\lambda)-G_2(\lambda_1))g,g) = \lambda(g,g)$. Thus $a \leq \lambda$. On the other hand $g = (G_1(\lambda_3)-G_1(\lambda_2))g$ yields by a similar argument that $a \geq \lambda_2$. Since $\lambda < \lambda_2$, this is a contradiction. It follows that the range of $G_2(\lambda)(1-G_1(\lambda))$ must contain only θ. Hence $G_2(\lambda) \leq G_1(\lambda)$.

In the preceding discussion, the commutativity of $G_1(\lambda)$ and $G_2(\mu)$ and the expressions for H, were sufficient for the result. result. It follows that a similar discussion will also show that $G_1(\lambda) \leq G_2(\lambda)$. Hence $G_1(\lambda) = G_2(\lambda)$.

§3

We consider in this section, a normal operator A. (Cf. Definition 2 of §4, Chapter VII). When we apply Theorem I of §1 above to A, we obtain as usual W, B and C with the usual properties (a) — (f). Since however $A^*A = AA^*$ we have $B = C$ and thus $A = WB = CW = BW$.

We also introduce $E = W^*W$, the projection on the initial set of W which is $[\mathfrak{R}_B] = \mathfrak{N}_B^\perp$. (Cf. Lemma 8 of §3, Chapter IV). Lemma 3 of §4, Chapter VII shows that WW^*, the projection on the final set of W is also E. Thus $(1-E)W = 0 = W(1-E)$. Taking adjoints we obtain $W^*(1-E) = 0 = (1-E)W^*$. Since E is the projection on $[\mathfrak{R}_B] = \mathfrak{N}_B^\perp$, we have $(1-E)B = B(1-E) = 0$.

Consider $\mathfrak{U} = 1-E+W$. Now $\mathfrak{U}\mathfrak{U}^* = (1-E+W)(1-E+W^*) = (1-E)+WW^*+W(1-E)+(1-E)W^* = 1-E+E+0+0 = 1$. Similarly $\mathfrak{U}\mathfrak{U}^* = 1$. Now $\mathfrak{U}\mathfrak{U}^* = 1$ shows that the range of $\mathfrak{U}$ is $\mathfrak{H}$ and $\mathfrak{U}^*\mathfrak{U} = 1$ implies that for every f and g, $(\mathfrak{U}f,\mathfrak{U}g) = (\mathfrak{U}^*\mathfrak{U}f,g) = (f,g)$. Definition 1 of §2, Chapter VI shows that $\mathfrak{U}$ is unitary.

We have then that $\mathfrak{U}B = (W+1-E)B = WB = A = BW = B(W+1-E) = B\mathfrak{U}$.

LEMMA 1. If A is normal there exists a unitary operator $\mathfrak{U}$ and a self-adjoint definite $B = \int_0^\infty \mu dF_1(\mu)$ such that $A = \mathfrak{U}B = B\mathfrak{U}$.

LEMMA 2. Let $\mathfrak{U}$ be unitary and let $\mathfrak{M}_1$, $\mathfrak{M}_{-1}$, E_1, E_{-1}, F and $\mathfrak{U}_1$ be as in Lemma 1 of Chapter VIII, §4. Then if D is a bounded self-adjoint operator which commutes with $\mathfrak{U}$, then D commutes with E_1, E_{-1}, F and $\mathfrak{U}_1$. Furthermore $D = E_1DE_1+E_{-1}DE_{-1}+FDF$, where FDF commutes with $\mathfrak{U}_1$, when both are regarded as contracted to the range of F_1.

If f is in $\mathfrak{M}_1$ then $\mathfrak{U}f = f$ and $Df = D\mathfrak{U}f = \mathfrak{U}Df$. Thus $f \in \mathfrak{M}_1$ implies $Df \in \mathfrak{M}_1$ and $DE_1 = E_1DE_1$. Taking adjoints yields $E_1D = E_1DE_1 = DE_1$. Similarly $E_{-1}D = E_{-1}DE_{-1} = DE_{-1}$. These results imply $FD = DF$ and $FD = F^2D = FDF$. Lemma 1 of §4, Chapter VIII implies

$$DE_1-DE_{-1}+D\mathfrak{U}_1 = D(E_1-E_{-1}+\mathfrak{U}_1) = D\mathfrak{U} = \mathfrak{U}D$$
$$=(E_1-E_{-1}+\mathfrak{U}_1)D = E_1D-E_{-1}D + \mathfrak{U}_1D.$$

Since $DE_1 = E_1D$, $DE_{-1} = E_{-1}D$, This implies $D\mathfrak{U}_1 = \mathfrak{U}_1D$. Inasmuch as $1 = E_1+E_{-1}+F$, $D = D(E_1+E_{-1}+F) = DE_1+DE_{-1}+DF = E_1DE_1+E_{-1}DE_{-1}+FDF$. The statement concerning the contraction of FDF is obvious.

LEMMA 3. Let $\mathfrak{U}$ be unitary and let $G(\phi)$ denote the resolution of the identity for $\mathfrak{U}$ given by Theorem II of §4, Chapter VIII. Then if D is a bounded self-adjoint transformation which commutes with $\mathfrak{U}$, then D commutes with $G(\phi)$ for $0 \leqq \phi \leqq 2\pi$.

PROOF: We know from Lemma 2 above that $D_1 = FDF$ commutes with $\mathfrak{U}_1$. Let us confine our attention to the range of F and then we may define $A_1 = \frac{1}{2}(\mathfrak{U}_1 + \mathfrak{U}_1^*)$, $B_1 = \frac{1}{2}i(\mathfrak{U}_1^* - \mathfrak{U}_1)$. Let W and C_1 be as in the proof of Lemma 3 of §4, Chapter VIII. D commutes with A_1 and B_1 by Lemma 1 of §4, Chapter VII and $E(\lambda)$ by Lemma 2 of §4, Chapter VII and with C by Corollary 2 to Lemma 3 of §4, Chapter VIII.

Lemma 1 of §4, Chapter VII shows that D_1 commutes with C_1^{-1}. Thus $D_1B_1C_1^{-1} = B_1D_1C_1^{-1} \subset B_1C_1^{-1}D_1$. Since $B_1C_1^{-1} = W$ on the range of C, we have $D_1Wf = WD_1f$ on the range of C and since this last set is dense and D_1 and W_1 are continuous, we obtain $D_1W = WD_1$. (Cf. Lemma 3 of §4, Chapter VIII)

Let $F_1 = \frac{1}{2}(1+W)$, $F_2 = \frac{1}{2}(1-W)$, as in the discussion following Lemma 4 of §4, Chapter VIII. Since D_1 commutes with W, D_1 commutes with F_1 and F_2. From the definition of $F_1(\phi)$ preceding the statement of Lemma 5 of §4, Chapter VIII, we see that D_1 must commute with $F_1(\phi)$ since D_1 commutes with F_1, F_2 and the $E(\phi)$.

We now return to the more general situation. We see firstly that $F(\phi)FD = F(\phi)FDF = F(\phi)D_1F = D_1F(\phi)F = DFF(\phi)F = D\cdot F(\phi)F$. Thus D commutes with $F(\phi)F$. We have shown above in Lemma 2 that D commutes with E_1 and E_{-1}. It follows from the definition of the $G(\phi)$ preceding Theorem II of §4, Chapter VIII, that D commutes with $G(\phi)$. This proves the Lemma.

Returning to the result of Lemma 1, we recall that $A = \mathfrak{U}B = B\mathfrak{U}$. Thus if $B = \int_0^\infty \mu dF_1(\mu)$, $\mathfrak{U}$ commutes with $F_1(\lambda)$. Now if $\mathfrak{U} = \int_0^{2\pi} e^{i\phi} dG_1(\phi)$, we see from Lemma 3 above that $F_1(\lambda)$ and $G_1(\phi)$ commute. Thus

$$A = \int_0^\infty \int_0^{2\pi} \rho e^{i\phi} dG_1(\phi) dF_1(\rho) = \int_0^{2\pi}\int_0^\infty \rho e^{i\phi} dF_1(\rho) dG_1(\phi).$$

This suggests that we may introduce the notion of a planar resolution of the identity and a corresponding planar integral. Our discussion will be simpler if we modify $G_1(\phi)$ by introducing $G_2(\phi) = G_1(\phi)+1-G_1(2\pi-0)$ for $0 \leq \phi < 2\pi$. It is readily seen that $\mathfrak{U} = \int_0^{2\pi} e^{i\phi} dG_2(\phi)+G_2(0)$, $G_2(2\pi-0) = 1$ and

$$A = \int_0^{2\pi}\int_0^\infty \rho e^{i\phi} dF_1(\rho) dG_2(\phi) + \int_0^\infty \rho dF_1(\rho) G_2(0).$$

If P is a point of the plane with polar coordinates (ρ,ϕ), $\rho \geq 0$, $0 \leq \phi < 2\pi$, (the origin has the coordinates $\rho = 0$, $\phi = 0$,) let $E_1(P) = F_1(\rho)G_2(\phi)$. Since $F_1(\rho)$ and $G_2(\phi)$ commute, one can show that $E_1(P)$ has the following properties:

(a): If P_1 and P_2 are two points with coordinates (ρ_1,ϕ_1), (ρ_2,ϕ_2) respectively and if Q is the point whose coordinates are $(\min(\rho_1,\rho_2), \min(\phi_1,\phi_2))$ then $E_1(P_1)\cdot E_1(P_2) = E_1(Q)$.

(b): If we let $E_1(\rho,\phi) = E_1(P(\rho,\phi))$ then $E_1(\rho+0,\phi+0) = E_1(\rho+0,\phi) = E_1(\rho,\phi+0) = E_1(\rho,\phi)$.

(c): If we now let $F_1(\rho) = E_1(\rho,2\pi-0)$, for $\rho > 0$, $F_1(0) = E_1(0,0)$ and $F_1(\rho) = 0$ if $\rho < 0$ then $F_1(\rho)$ is a resolution of the identity. Similarly if we let $G_2(\phi) = \lim_{\rho\to\infty} E_1(\rho,\phi)$ for $0 \leq \phi < 2\pi$, $G_2(\phi) = 0$ for $\phi < 0$, $G_2(\phi) = 1$ for $\phi \geq 2\pi$, then $G_2(\phi)$ is a resolution of the identity with $G_2(2\pi-0) = 1$.*

DEFINITION 1. A family of projections $E_1(P)$ having the properties (a), (b) and (c) above, will be called a planar resolution of the identity.

Let us consider a sector S of a circular ring, S not containing the polar axis. Thus S is the set of points (ρ,ϕ) with $\rho_1 < \rho \leq \rho_2$, $\phi_1 < \phi \leq \phi_2$. We can associate with it a projection $E_1(S) = (F_1(\rho_2)-F_1(\rho_1))(G_2(\phi_2)-G_2(\phi_1)) = E_1(P_1)-E_1(P_2)-E_1(P_3)+E_1(P_4)$, where P_1, P_2, P_3 and P_4 are the points

---

* These statements are redundant. For instance in (c) it is only necessary to show that $\lim_{\rho\to\infty} F_1(\rho) = 1$ and that $G_2(2\pi-0) = 1$, since the other properties are consequences of (a); (b) and the definitions of $F_1(\rho)$ and $G_2(\phi)$. For example a consideration of projections associated with the areas as in the following discussion will show that $G_2(\phi+0) = G_2(\phi)$.

(ρ_2,ϕ_2), (ρ_2,ϕ_1), (ρ_1,ϕ_2) and (ρ_1,ϕ_1) respectively. ($E_1(S)$ is a projection by Lemma 4 of §1, Chapter VI).

Any such S can be expressed as the logical sum of mutually exclusive smaller S_α, i.e. $S = S_1 \cup \dots \cup S_n$, and we will call this a partition of S. We can then form for any function ψ, the partial sums $\Sigma\, \psi(Q_\alpha)E(S_\alpha)$ where $Q_\alpha \in S_\alpha$. If ψ is continuous, it can be shown that for every sequence of partitions such that the maximum diameter of the S_α's approaches zero, these partial sums approach a limit which we will denote:

$$\iint_S \psi(P)dE_1(P).$$

Furthermore a familiar discussion will show that

$$\iint_S \psi(P)dE_1(P) = \int_{\rho_1}^{\rho_2}\int_{\phi_1}^{\phi_2} \psi(P(\rho,\phi))dG_2(\phi)dF_1(\rho).$$

If we close S and call the result S', we define

$$E_1(S') = (F_1(\rho_2)-F_1(\rho_1-0))(G_2(\phi_2)-G_2(\phi_1-0)).$$

We can partition S' into closed and partially closed sectors and again obtain for every continuous $\psi(P)$,

$$\iint_{S'} \psi(P)dE_1(P).$$

This equals

$$\int_{\rho_1-0}^{\rho_2}\int_{\phi_1-0}^{\phi_2} (P(\rho,\phi))dG_2(\phi)dF_1(\rho).$$

One may also introduce the notion of an improper integral, so that if S_0 denotes the entire plane, we may define the planar integral over S_0 and obtain:

$$\iint_{S_0} \psi(P)dE_1(P) = \int_{0-0}^{\infty}\int_{0-0}^{2\pi} \psi(P(\rho,\phi))dG_2(\phi)dF_1(\rho)$$

$$= \int_0^\infty\int_0^{2\pi} \psi(P(\rho,\phi))dG_2(\phi)dF_1(\rho) + \int_0^\infty \psi(P(\rho,0))dF_1(\rho)dG_2(0) +$$

$$\psi(P(0,0))G_2(0)F_1(0).$$

In particular if we let $\psi(P(\rho,\phi)) = z(P) = \rho e^{i\phi}$, we obtain:

$$A = \iint_{S_0} z dE_1(P).$$

THEOREM III. If A is a normal operator, there exists a planar resolution of the identity $E_1(P)$ such that

$$A = \iint_{S_0} z dE_1(P).$$

Lemma 3 above can also be used to show

LEMMA 4. If $\mathfrak{U}$ is unitary, the equations $\mathfrak{U} = \int_0^{2\pi} e^{i\phi} dG(\phi)$, $G(0) = 0$, determine the resolution of the identity $G(\phi)$ uniquely.

Let G be as in Theorem II of §4, Chapter VIII and suppose that we also have $\mathfrak{U} = \int_0^{2\pi} e^{i\phi} dG_1(\phi)$, $G_1(0) = 0$.

Suppose now that for $0 < \phi < 2\pi$, $G_1(\phi)$ commutes with $\mathfrak{U}$ and hence by Lemma 3 above, $G_1(\phi)$ commutes with $G(\psi)$. We will prove that $G_1(\phi)(1-G(\phi_1)) = 0$. Lemma 4 of §1, Chapter VI shows that $G_1(\phi)(1-G(\phi))$ is a projection. Now suppose that there is a $f \neq \theta$ such that $f = G_1(\phi)(1-G(\phi))f$. Since $G_1(0+0) = G_1(0) = 0$ and $G(\phi+0) = G(\phi)$, we can find an α and a ϕ_1 such that $g = (G_1(\phi)-G_1(\alpha))(1-G(\phi))f \neq \theta$ and $0 < \alpha < \phi < \phi_1$.

Since $G_1(\phi)$ and $G(\psi)$ commute, we have that $g = (G_1(\phi)-G_1(\alpha))g = (1-G(\phi_1))g$. Let a be defined as the equation $a(g,g) = (\mathfrak{U}g,g)$ which is possible since $g \neq \theta$. The argument of a is the same as that of $(\mathfrak{U}g,g) = (\mathfrak{U}(G_1(\phi)-G_1(\alpha))g,g) = \int_\alpha^\phi e^{i\phi} d(G_1(\phi)g,g)$. A consideration of the partial sums will show that their arguments always lie between ϕ and α and it follows that the argument of their limit $(\mathfrak{U}g,g)$ is in this interval closed. Thus the argument of a is $\leqq \phi$.

However we also have $(1-G(\phi_1))g = g$ and $\mathfrak{U} = \int_0^{2\pi} e^{i\phi} dG(\phi)$ and a similar argument will show that the argument of a is $\geqq \phi_1$. Since $\phi < \phi_1$, this yields a contradiction. Thus f in the range of $G_1(\phi)(1-G(\phi))$ implies $f = \theta$ and hence $G_1(\phi) \leqq G(\phi)$.

In this discussion, we have used simply that $G_1(\phi)$ and $G(\psi)$ commute, $G_1(0) = 0$ and the two expressions for $\mathfrak{U}$. A similar argument will therefore show that $G(\phi) \leqq G_1(\phi)$ and hence $G_1(\phi) = G(\phi)$. Since $G(0) = 0$, we now have $G_1(\phi) = G(\phi)$ for $0 \leqq \phi < 2\pi$. But $G_1(2\pi)$ must be 1. For if f is in the range of $1-G_1(2\pi)$, $\mathfrak{U}f = \int_0^{2\pi} e^{i\phi} dG_1(\phi)f = \theta$ and $0 = |\mathfrak{U}f| = |f|$. Thus $1-G_1(2\pi) = 0$ and $G_1(2\pi) = G(2\pi)$. This completes the proof of the Lemma.

We may also establish.

LEMMA 5. If A is c.a.d.d. and F is a projection such that $FA \subset AF$, then F commutes with B, C, W, $W^*W = E_1$, $WW^* = E_2$, where B, C, and W are as in

Theorem I of §1 above. In particular if A is normal and equals $\mathfrak{U}B = B\mathfrak{U}$ where $\mathfrak{U}$ is unitary, then F commutes with $\mathfrak{U}$, B, $F_1(\rho)$, $G_2(\phi)$ and $E_1(P)$ discussed in the proof of Theorem III above.

By the Corollary to Theorem V of §2, Chapter IV, $FA \subset AF$ implies $A^*F = (FA)^* \supset (AF)^* \supset FA^*$. Thus $FB^2 = FA^*A \subset A^*FA \subset A^*AF = B^2F$, or F commutes with B^2. Lemma 2 of §1, above shows that F commutes with B.

A discussion similar to that given in the proof of Corollary 3 to Theorem II of §2 above, now shows that F commutes with E_1 and W. Since A and A^* are now interchangeable, we obtain that F also commutes with C, E_2 and W^*.

For a normal A, F commutes with $\mathfrak{U} = W+1-E$. Lemma 3 above and Corollary 3 to Theorem II of §2, above shows that F commutes with $G_1(\phi)$ and $F_1(\rho)$. Since $G_2(\phi) = G_1(\phi)+1-G_1(2\pi-0)$, F also commutes with $G_2(\phi)$. Since $E_1(P) = G_2(\phi)F_1(\rho)$, F commutes with $E_1(P)$.

We conclude our discussion of normal operators by showing

COROLLARY 1. The planar resolution of the identity $E_1(P)$ of Theorem III above is unique.

Let $F_1(\rho)$ and $G_2(\phi)$ be as in property (c) of Definition 1 above. A proof similar to that of Corollary 4 of the preceding section will show that $F_1(\rho)$ depends only on A. Similarly, a discussion similar to the proof of Lemma 4 above will show that $G_2(\phi)$ is unique. Property (a) of Definition 1 above can now be used to show that $E_1(P)$ is determined.

COROLLARY 2. $A^* = \int_{S_0} \bar{z}\,dE_1(P)$.

This is a consequenc of $A^* = B\mathfrak{U}^* = B\mathfrak{U}^{-1}$. For if $\mathfrak{U} = \int_0^{2\pi} e^{i\phi} dE(\phi)$ and $V = \int_0^{2\pi} e^{-i\phi} dE(\phi)$, we have by Lemma 5 of §2, Chapter VII, $\mathfrak{U}V = V\mathfrak{U} = 1$ and hence $\mathfrak{U}^{-1} = V$. If we apply the discussion preceding Theorem III above to $A^* = VB$, instead of $A = \mathfrak{U}B$, we will get the result stated in the Corollary.

CHAPTER X

SYMMETRIC OPERATORS

In this Chapter, we discuss symmetric transformations, i.e. those for which $H \subset H^*$. (Cf. Definition 1 of §3, Chapter IV). We will be concerned in particular with the notions of a symmetric extension and of maximality and its relationship with the property of being self-adjoint. (Cf. Definition 3 of §3, Chapter IV).

§1

In this section, H will be considered to be closed and symmetric.

LEMMA 1. If f and g are in the domain of H, then $((H+i)f,(H+i)g) = ((H-i)g,(H-i)g) = (Hf,Hg)+(f,g)$.

Expanding the expressions, will yield this result since $(f,Hg) = (Hf,g)$. (Cf. Definition 1 of §3, Chapter IV).

LEMMA 2. (a): The set $\mathfrak{D}$ of pairs $\{(H+i)f,(H-i)f\}$ for f in the domain of H is the graph of a transformation V. (b): V is closed and isometric. (c): $(H+i)^{-1}$ exists and $V = (H-i)(H+i)^{-1}$. (d): If V_1, V_2 correspond to H_1, H_2 respectively as in the above, then H_1 is a (proper) symmetric extension of H_2 if and only if V_1 is a (proper) isometric extension of V_2. (e): If $\mathfrak{M}_i$ is the set of $f \in \mathfrak{H}$ such that $H^*f = if$, the domain of V is $\mathfrak{M}_i^\perp$. (f): If $\mathfrak{M}_{-i}$ is the set of $f \in \mathfrak{H}$ such that $H^*f = -if$, the range of V is $\mathfrak{M}_{-i}^\perp$. (g): The range of $V-1$ is dense. (h): $(V-1)^{-1}$ exists and $H = -i(V+1)(V-1)^{-1}$. (i): If E is a projection such that $EV \subset VE$, then $EH \subset HE$.

Proof of (a). If f is such that $(H+i)f = \theta$ then $0 = |(H+i)f|^2 = ((H+i)f,(H+i)f) = (Hf,Hf)+(f,f) = |Hf|^2 + |f|^2$ by Lemma 1 above. Thus $(H+i)f = \theta$ implies $|f|^2 = 0$ and $f = \theta$. Hence $(H-i)f = \theta$. Thus $\{\theta,h\} \in \mathfrak{D}$ implies $h = \theta$. $\mathfrak{D}$ is easily seen to be additive and hence Lemma 3 of §1, Chapter IV shows that $\mathfrak{D}$ is the graph of a transformation.

Proof of (b). We first show that inasmuch as H is closed, $\mathfrak{D}$ is closed. Let $\{\phi,\psi\}$ be a pair in the closure of $\mathfrak{D}$. Let $[\{(H+i)f_n,(H-i)f_n\}]$ be a sequence in $\mathfrak{D}$ such that $(H+i)f_n \longrightarrow \phi$, $(H-i)f_n \longrightarrow \psi$. Let $f = \frac{1}{2}i(\psi-\phi) = \frac{1}{2}i(\lim (H-i)f_n - \lim (H+i)f_n) = \frac{1}{2}i \lim (-2if_n) = \lim f_n$. Similiarly $f^* = \frac{1}{2}(\phi+\psi) = \lim Hf_n$. Thus if we let $f = \frac{1}{2}i(\psi-\phi)$ then Hf exists and $= \frac{1}{2}(\phi+\psi)$, since H is closed. We also have $if = \frac{1}{2}(\phi-\psi)$, and thus $(H+i)f = \phi$, $(H-i)f = \psi$. Hence $\{\phi,\psi\}$ is in $\mathfrak{D}$. Thus $\mathfrak{D}$ contains its limit points and thus V is closed. It follows by definition that V is closed.

Since $\mathfrak{D}$ is a linear manifold, V is additive. Furthermore, if ϕ_1 and ϕ_2 are in the domain of V then $\phi_1 = (H+i)f_1$, $\phi_2 = (H+i)f_2$, $V\phi_1 = (H-i)f_1$ and $V\phi_2 = (H-i)f_2$ for some f_1 and f_2 in the domain of H. Hence Lemma 1 above implies $(\phi_1,\phi_2) = ((H+i)f_1,(H+i)f_2) = ((H-i)f_1,(H-i)f_2) = (V\phi_1, V\phi_2)$. Thus Definition 2 of §2, Chapter VI, shows that V is isometric.

Proof of (c). In the proof of (a) above, we have shown that $(H+i)f = \theta$ implies $f = \theta$. Lemma 4 and Definition 2 of §1, Chapter IV, now show that $(H+i)^{-1}$ exists. Now if ϕ is in the domain of V, $\phi = (H+i)f$ for f in the domain of H. Hence $(H+i)^{-1}\phi$ exists and equals f. Also $V\phi = (H-i)f = (H-i)(H+i)^{-1}\phi$. Thus $V \subset (H-i)(H+i)^{-1}$. On the other hand, if ϕ is in the domain of $(H-i)(H+i)^{-1}$, we let $f = (H+i)^{-1}\phi$ and $\psi = (H-i)(H+i)^{-1}\phi$. Since $\phi = (H+i)f$, $\psi = (H-i)f$, we have $V\phi = \psi$. This shows that $V \supset (H-i)(H+i)^{-1}$ and with our previous inclusion proves the equality.

H_1 a proper extension of H_2 is equivalent to H_1+i a proper extension of H_2+i, which in turn is equivalent to $(H_1+i)^{-1}$ being a proper extension of $(H_2+i)^{-1}$. Now the domain of $V = (H-i)(H+i)^{-1}$ is precisely the domain of $(H+i)^{-1}$, since $(H-i)f$ is defined on the range of $(H+i)^{-1}$. Hence

$(H+i)^{-1}$ being a proper extension of $(H_2+i)^{-1}$ is equivalent to V_1 being a proper extension of V_2. These equivalences are sufficient to prove (d).

Proof of (e). Lemma 4 of §2, Chapter VI shows that since V is closed, both its domain $\mathfrak{D}_V$ and range $\mathfrak{R}_V$ are closed linear manifolds. Now $\mathfrak{D}_V$ is also the domain of $(H+i)^{-1}$ which is the range of $H+i$. But $\mathfrak{D}_V = \mathfrak{R}_{H+i}$ must be the orthogonal complement of the zeros of $(H+i)^*$ by Theorem VI of §2, Chapter IV. Now, by Theorem V of §2, Chapter IV, $(H+i)^* \supset H^*-i$ and $H^* = (H+i-i)^* \supset (H+i)^*+i$ or $H^*-i \supset (H+i)^*$. It follows that $(H+i)^* = H^*-i$. Thus $\mathfrak{D}_V$ is the orthogonal complement of the set for which $(H^*-i)f = \theta$ or for which $H^*f = if$.

(f) is proven in a similar way.

Proof of (g). If f is in the domain of H, there is a ϕ in the domain of V , such that $(H+i)f = \phi$, $(H-i)f = V\phi$. Subtracting, we get $2if = (1-V)\phi$ or $f = \frac{1}{2}i(V-1)\phi$. Thus the range of $\frac{1}{2}i(V-1)$ includes the domain of H, which is dense. The statement (g) follows easily from this.

Proof of (h). We first prove that if V is isometric and $\mathfrak{R}_{1-V}$ is dense, then $(1-V)^{-1}$ exists. Now $(V-1)^{-1}$ exists if and only if $(V-1)\phi = \theta$ implies $\phi = \theta$. (Cf. Lemma 4 of §1, Chapter IV). Let us suppose that $(V-1)\phi = \theta$ or $V\phi = \phi$. For ψ in the domain of V, we have

$$0 = (V\phi,V\psi)-(\phi,\psi) = (\phi,V\psi)-(\phi,\psi) = (\phi,V\psi-\psi).$$

Thus ϕ is orthogonal to $\mathfrak{R}_{V-1}$ and since this last set is dense, we must have $\phi = \theta$. Hence $(V-1)\phi = \theta$ implies $\phi = \theta$.

Furthermore if f is in the domain of H, we have for a $\phi \in \mathfrak{D}_V$, $-2if = (V-1)\phi$ and $2Hf = (V+1)\phi$. It follows that $\phi = -2i(V-1)f$ and $Hf = -i(V+1)(V-1)^{-1}f$. Thus $H \subset -i(V+1)(V-1)^{-1}$. On the other hand, in the above, we have shown that if $g = (V-1)\phi$, g is in the domain of H. Thus $\mathfrak{R}_{V-1} \subset \mathfrak{D}_H$. This is equivalent to $\mathfrak{D}_{(V-1)^{-1}} = \mathfrak{R}_{V-1} \subset \mathfrak{D}_H$. If $T = -i(V+1)(V-1)^{-1}$, $\mathfrak{D}_T = \mathfrak{D}_{(V-1)^{-1}}$ since $V+1$ is defined everywhere on the range of $(V-1)^{-1}$. Thus $\mathfrak{D}_T = \mathfrak{D}_{(V-1)^{-1}} \subset \mathfrak{D}_H$, and with our previous inclusion $H \subset T$, this shows $T = H$.

Proof of (k). If $EV \subset VE$, we see that the domain of these transformations include $\mathfrak{D}_V$. We also have $E(V-1) \subset (V-1)E$. If f is in the range of $V-1$, i.e. $f = (V-1)\phi$, for a ϕ in the domain of V, then $Ef = E(V-1)\phi = (V-1)E\phi$, and Ef is also in the range of $(V-1)$. Furthermore $Ef = (V-1)E\phi = (V-1)E(V-1)^{-1}f$, or $(V-1)^{-1}Ef = E(V-1)^{-1}f$. Since this holds for every f in $\mathfrak{R}_{V-1} = \mathfrak{D}_{(V-1)^{-1}}$, we must have $E(V-1)^{-1} \subset (V-1)^{-1}E$.

We also have $E(V+1) \subset (V+1)E$. Hence $E(V+1)(V-1)^{-1} \subset (V+1)E(V-1)^{-1} \subset (V+1)(V-1)^{-1}E$. The expression for H obtained in the above now shows that $EH \subset HE$.

DEFINITION 1. If H is closed symmetric and V is as in (a) of Lemma 2 above, then $V = V_H$ is called the Cayley transform of H.

In the proof of (h) above, we have shown:

LEMMA 3. If V is isometric and such that $\mathfrak{R}_{V-1}$ is dense, then $(V-1)^{-1}$ exists.

LEMMA 4. Let V be closed and isometric and such that $\mathfrak{R}_{V-1}$ is dense. Lemma 3 above shows that $(V-1)^{-1}$ exists. Let $H = -i(V+1)(V-1)^{-1}$. Then (a): H is closed symmetric. (b): The Cayley transform of H is V. (c): Let $\mathfrak{N}_i = \mathfrak{D}_V^\perp$, $\mathfrak{N}_{-i} = \mathfrak{R}_V^\perp$, then the domain of H and $\mathfrak{N}_i$ or $\mathfrak{N}_{-i}$ have only θ in common. (d): The domain of H^* consists of elements in the form $f+g_1+g_2$ where $f \in \mathfrak{D}_H$, $g_1 \in \mathfrak{N}_{-i}$, $g_2 \in \mathfrak{N}_i$ and $H^*(f+g_1+g_2) = Hf-ig_1+ig_2$.

PROOF OF (a): If ϕ is in the domain of V, let $f = \frac{1}{2}i(V-1)\phi$. Then $\phi = -2i(V-1)f$, and $Hf = (V+1)(-i(V-1)^{-1})f = \frac{1}{2}(V+1)\phi$. Thus if f_1 and f_2 are in the domain of H and ϕ_1 and ϕ_2 denote the corresponding ϕ's, we have $(Hf_1,f_2) = ((V+1)\phi_1, i(V-1)\phi_2) = -i((V+1)\phi_1,(V-1)\phi_2) = -i[(V\phi_1,V\phi_2)+(\phi_1,V\phi_2)-(V\phi_1,\phi_2)-(\phi_1,\phi_2)] = i[(V\phi_1,\phi_2)-(\phi_2,V\phi_1)]$ because for an isometric V, $(V\phi_1,V\phi_2) = (\phi_1,\phi_2)$. Similarly $(f,Hf_2) = i[(V\phi_1,\phi_2)-(\phi_1,V\phi_2)]$. Thus for every f_1 and f_2 in $\mathfrak{D}_H$, $(f_1,Hf_2) = (Hf_1,f_2)$. Furthermore $\mathfrak{D}_H = \mathfrak{D}_{(V-1)^{-1}}$ is by hypothesis dense. Thus Definition 1 of §3, Chapter IV shows that H is symmetric.

The proof that H is closed is analgous to the proof of the closure of V, in (b) of Lemma 2 above.

PROOF OF (b): If ϕ is in the domain of V, $f = \frac{1}{2}i(V-1)\phi$ is in the domain of H with $Hf = \frac{1}{2}(V+1)\phi$. Thus $\phi = \frac{1}{2}(V+1)\phi - \frac{1}{2}(V-1)\phi = (H+i)f$ and $V\phi = \frac{1}{2}(V+1)\phi + \frac{1}{2}(V-1)\phi = (H-i)f$. Thus V is included in the Cayley transform of H. If, however, V_H were a proper extension of V, $(V_H-1)^{-1}$ would be a proper extension of $(V-1)^{-1}$ since $\mathfrak{R}_{(V-1)^{-1}} = \mathfrak{D}_V$. Hence $\mathfrak{D}_{(V-1)^{-1}}$ is included in but not equal to $\mathfrak{D}_{(V_H-1)^{-1}}$. However we see from our hypotheses and (a) of Lemma 2 above that these sets are both $\mathfrak{D}_H$. This is a contradiction and we have $V = V_H$.

PROOF OF (c). Let us suppose that g is in $\mathfrak{N}_1 \cdot \mathfrak{D}_H$ and $g \neq \theta$. Then $g = i(V-1)\phi$ for $\phi \in \mathfrak{D}_V$. Since $\mathfrak{N}_1 = \mathfrak{D}_V^\perp$, we must have $0 = (g,\phi) = i(V\phi-\phi,\phi)$. This implies $(V\phi,\phi) = (\phi,\phi) = |\phi|^2 = |V\phi|\cdot|\phi|$. This is only possible if $V\phi = k\phi$ for a constant k. If $g \neq \theta$, $\phi \neq \theta$ and $(V\phi,\phi) = (\phi,\phi)$ implies $k = 1$. Thus $V\phi = \phi$ or $(V-1)\phi = \theta$. This implies $g = i(V-1)\phi = \theta$ contrary to our supposition. Thus $g = \theta$ and $\mathfrak{N}_1 \cdot \mathfrak{D}_H = \{\theta\}$. The proof of $\mathfrak{N}_{-1} \cdot \mathfrak{D}_H = \{\theta\}$ is similar.

PROOF OF (d). Let $\mathfrak{K}$ denote the graph of H and $\mathfrak{K}^*$ denote the set of pairs $\{f,H^*f\}$, i.e. the graph of H^*. Since H is symmetric, we have $\mathfrak{K} \subset \mathfrak{K}^*$. Consider $\mathfrak{K}^\perp \cdot \mathfrak{K}^*$ and let us suppose that $\{h,H^*h\}$ is in $\mathfrak{K}^\perp \cdot \mathfrak{K}^*$. We have for every f in the domain of H,

$$(f,H^*h) = (Hf,h),$$
$$0 = (\{f,Hf\},\{h,H^*h\}) = (f,h)+(Hf,H^*f).$$

This implies that for every ϕ in the domain of V,

$$(\tfrac{1}{2}i(V-1)\phi,H^*h)-(\tfrac{1}{2}(V+1)\phi,h) = 0,$$
$$(\tfrac{1}{2}i(V-1)\phi,h)+(\tfrac{1}{2}(V+1)\phi,H^*h) = 0.$$

A simple calculation will show that these equations are equivalent respectively to

$$(V\phi,H^*h-ih)-(\phi,H^*h+ih) = 0,$$
$$(V\phi,H^*h-ih)+(\phi,H^*h+ih) = 0,$$

and these equations are equivalent to

$$(V\phi,H^*h-ih) = 0,$$
$$(\phi, H^*h+ih) = 0.$$

Thus if $-2ig_1 = H^*h-ih$, $2ig_2 = H^*h+ih$, g_1 is in $\mathfrak{N}_{-1} = \mathfrak{R}_V^\perp$, g_2 is in $\mathfrak{N}_1 = \mathfrak{D}_V^\perp$, $h = g_1+g_2$, $H^*h = -ig_1+ig_2$. Thus if $\{h,H^*h\} \in \mathfrak{K}^*\mathfrak{K}$, $h = g_1+g_2$, $g_1 \in \mathfrak{N}_{-1}$, $g_2 \in \mathfrak{N}_1$, $H^*h = -ig_1+ig_2$.

On the other hand if $H = g_1+g_2$, $h^* = -ig+ig_2$, reversing the above discussion, will show that $g_1 \in \mathfrak{N}_{-i}$ and $g_2 \in \mathfrak{N}_i$ implies

$$(f,h^*) = (Hf,h),$$

$$0 = (\{f,Hf\},\{h,h^*\}),$$

for every f in the domain of H. Theorem II of §2, Chapter IV, shows that H^*h exists and equals h^*. Thus we may conclude that $\{h,h^*\} \in \mathfrak{H}^*\mathfrak{H}^\perp$.

If k is in the domain of H^*, $\{k,H^*k\}$ is in $\mathfrak{H}^*$ and $\{k,H^*k\} = \{f,Hf\}+\{h,H^*h\}$ where $\{f,Hf\} \in \mathfrak{H}$, $\{h,H^*h\} \in \mathfrak{H}^\perp.\mathfrak{H}^*$, by Corollary 1 to Theorem VI of §5, Chapter II. From the above, we obtain $k = f+h = f+g_1+g_2$ where $f \in \mathfrak{D}_H$, $g_1 \in \mathfrak{N}_{-i}$, $g_2 \in \mathfrak{N}_i$, and $H^*k = Hf+H^*h = Hf-ig_1+ig_2$. Thus every element k in $\mathfrak{D}_{H^*}$ is in the desired form and the converse is also readily shown when our previous results are used. Furthermore, the given formula for H^*k holds.

This completes the proof of the Lemma. We may now state:

THEOREM I. If H is closed symmetric, there exists a closed isometric V_H called the Cayley Transform, having the properties (a) to (i) of Lemma 2 above. If V is closed isometric and such that $\mathfrak{R}_{V-1}$ is dense, then there exists a symmetric H having properties (a) to (d) above.

COROLLARY 1. A closed symmetric H is self-adjoint if and only if V_H is unitary, i.e. $\mathfrak{D}_V^\perp = \mathfrak{R}_V^\perp = \{\theta\}$.

If V_H is unitary and $\mathfrak{D}_V^\perp = \{\theta\}$, $\mathfrak{R}_V^\perp = \{\theta\}$, (d) of Lemma 3 shows that the domain of H^* is simply that of H. Since $H \in H^*$, we must have $H = H^*$.

If V_H is not unitary, either $\mathfrak{D}_V^\perp$ or $\mathfrak{R}_V^\perp \neq \{\theta\}$. Let us suppose that $g_1 \neq \theta$ is in $\mathfrak{D}_V^\perp$. By (c) of Lemma 2 above, g_1 is not in $\mathfrak{D}_H$. However (d) of Lemma 2 above shows that g_1 is in $\mathfrak{D}_{H^*}$. Thus $H^* \neq H$.

COROLLARY 2. If H is closed symmetric, H has a maximal symmetric extension. (Cf. Definition 3 of §3, Chapter IV). H has a closed self-adjoint extension if and only if $\mathfrak{D}_V^\perp$ has the same dimensionality as $\mathfrak{R}_V^\perp$.

PROOF: If V_H is such that either $\mathfrak{R}_V^\perp$ or $\mathfrak{D}_V^\perp = \{\theta\}$, then V_H has no isometric extension and it follows from (d) of Lemma 2 above that H is maximal symmetric. Thus we may consider the case where both $\mathfrak{R}_V^\perp$ and $\mathfrak{D}_V^\perp$ are not $\{\theta\}$. For convenience let us assume that $\mathfrak{D}_V^\perp$ has dimensionality less than or equal to $\mathfrak{R}_V^\perp$. By using Lemma 7 of §2, Chapter VI, we can find an isometric V' with domain $\mathfrak{D}_V^\perp$ and range included in $\mathfrak{R}_V^\perp$. Lemma 8 of §2, Chapter VI shows that $V_1 = V \oplus V'$ is an isometric transformation such that $V_1 \supset V$, $\mathfrak{D}_{V_1} = \mathfrak{H}$. Since $V_1 \supset V$, $\mathfrak{R}_{V_1-1} \supset \mathfrak{R}_{V-1}$. Since the latter is dense, $\mathfrak{R}_{V_1-1}$ is also and Lemma 4 above shows that there is a symmetric H_1, whose Cayley transform is V_1. Since $\mathfrak{D}_{V_1} = \mathfrak{H}$, H_1 must be maximal symmetric as we remarked above. Since $V_1 \supset V$, Lemma 2 (d) above shows that H_1 is a proper symmetric extension of H. A similar argument holds if the dimensionality of $\mathfrak{D}_V^\perp$ is greater than that of $\mathfrak{R}_V^\perp$, with however the result that $\mathfrak{R}_{V_1} = \mathfrak{H}$.

V has a unitary extension V_1, if and only if the dimensionality of $\mathfrak{R}_V^\perp$ is the same as that of $\mathfrak{D}_V^\perp$. (Cf. Lemma 10 of §2, Chapter VI). Since $\mathfrak{R}_{V-1}$ is dense, we see from Corollary 1 above, that H_1 can be taken as self-adjoint if and only if the dimensionality of $\mathfrak{R}_V^\perp$ equals that of $\mathfrak{D}_V^\perp$.

We have also shown:

COROLLARY 3. H is maximal symmetric if and only if at least one of the $\mathfrak{D}_V^\perp$ or $\mathfrak{R}_V^\perp$ consists of θ alone.

§2

In this section, we present an analysis of maximal symmetric operators, obtaining both structural and existential results.

DEFINITION 1. Let $\phi_0, \phi_1, \phi_2, \ldots\ldots$ be a complete orthonormal set in $\mathfrak{H}$. (Cf. the end of §6, Chapter II.) Let V_0 be the transformation defined by the equation, $V_0(\Sigma_{\alpha=0}^{\infty} a_\alpha \phi_\alpha) = \Sigma_{\alpha=0}^{\infty} a_\alpha \phi_{\alpha+1}$. Let E denote the projection on $\mathfrak{M}_0 = \mathfrak{M}(\{\phi_1, \phi_2, \ldots\})$, i.e. the range of V_0. (Cf. §6, Chapter II, Theorem XI).

LEMMA 1 (a): V_0 is isometric with domain $\mathfrak{H}$ and range $\mathfrak{M}_0$: (b): V_0 is partially isometric with initial set $\mathfrak{H}$

and final set $\mathfrak{M}_\varnothing$. $V_0^* = V_0^{-1}E$, $V_0^*V_0 = V_0^{-1}V_0 = 1$, $V_0V_0^* = E$: (c): $\mathfrak{R}_{V-1}$ and $\mathfrak{R}_{V-1-1}$ are dense: (d): $(V_0-1)^{-1}$ and $(V_0^{-1}-1)^{-1}$ exist and $(V_0^{-1}+1)(V_0^{-1}-1)^{-1} = -(V_0+1)\cdot(V_0-1)^{-1}$.

(a): is a consequence of Lemma 7 of Chapter VI §2.

(b): follows from Definition 1 and Lemma 1 of §3, Chapter VI.

(c): We first show that $\mathfrak{R}_{V_0-1}$ is dense. By Theorem VI of §2, Chapter IV, we have $(\mathfrak{R}_{V_0-1})^\perp = \mathfrak{N}_{V_0^*-1}$. Thus if $f \in (\mathfrak{R}_{V_0-1})^\perp$, $(V_0^*-1)f = \theta$ or $V_0^*f = f$. Since $f \in \mathfrak{H}$, $f = a_0\phi_0 + a_1\phi_1 + \ldots\ldots$ (Cf. Theorem XII of §6, Chapter II). We have $V_0^*f = V_0^{-1}E_0(a_0\phi_0+a_1\phi_1+\ldots) = V_0^{-1}(a_1\phi_1+a_2\phi_2+\ldots) = a_1\phi_0+a_2\phi_1+\ldots.$ Thus $V_0^*f = f$ implies $a_0 = a_1$, $a_1 = a_2, \ldots$ etc. Since $\Sigma_{\alpha=0}^{\infty}|a_\alpha|^2 < \infty$, $\lim_{\alpha\to\infty} a_\alpha = 0$, and hence $a_\alpha = 0$ for $\alpha = 0, 1, \ldots$. Hence $f = \theta$. Thus $f \in (\mathfrak{R}_{V_0-1})^\perp$ implies $f = \theta$ and hence $\mathfrak{R}_{V_0-1}$ is dense.

$\mathfrak{R}_{V_0^{-1}-1}$ is also dense. Consider $\mathfrak{R}_{V_0^*-E}$ and suppose $g \in \mathfrak{R}_{V_0^*-E}$. Then $g = (V_0^*-E)f = (V_0^{-1}E-E)f = (V_0^{-1}-1)Ef$. Thus $g \in \mathfrak{R}_{V_0^*-E}$ implies $g \in \mathfrak{R}_{V_0^{-1}-1}$ or $\mathfrak{R}_{V^*-E} \subset \mathfrak{R}_{V_0^{-1}-1}$. Hence $(\mathfrak{R}_{V_0^{-1}-1})^\perp \subset (\mathfrak{R}_{V_0^*-E_0})^\perp = \mathfrak{N}_{V_0-E}$ by Theorem VI of §2, Chapter IV. Thus $f \in (\mathfrak{R}_{V_0^{-1}-1})^\perp$ implies $f \in \mathfrak{N}_{V_0-E}$ or $(V_0-E)f = \theta$ or $V_0f = Ef$. Letting $f = a_0\phi_0+a_1\phi_1+\ldots$ as before we obtain $V_0f = a_0\phi_1+a_1\phi_2+\ldots$ while $Ef = a_1\phi_1+a_2\phi_2+\ldots$. Since $V_0f = Ef$ we must have $a_0 = a_1$, $a_1 = a_2$, ... etc. Thus $\Sigma_\alpha|a_\alpha|^2 < \infty$ again implies $a = 0$, $\alpha = 0, 1, \ldots$ and $f = \theta$. Thus $f \in (\mathfrak{R}_{V_0^{-1}-1})^\perp$ implies $f = \theta$ and $\mathfrak{R}_{V_0^{-1}-1}$ is dense.

(d): Lemma 3 of §1 above shows that $(V_0-1)^{-1}$ and $(V_0^{-1}-1)^{-1}$ exist. Now $V_0^{-1}-1$ is defined only on $\mathfrak{M}$ and thus $V_0^{-1}-1 = V_0^{-1}-V_0V_0^{-1} = (1-V_0)V_0^{-1}$ since $V_0V_0^{-1} = 1$ on $\mathfrak{M}$. Thus $(V_0^{-1}-1)^{-1} = ((1-V_0)V_0^{-1})^{-1} = V_0(1-V_0)^{-1}$, and $(V_0^{-1}+1)(V_0^{-1}-1)^{-1} = (V_0^{-1}+1)V_0(1-V_0)^{-1} = (1+V_0)(1-V_0)^{-1} = -(V_0+1)\cdot(V_0-1)^{-1}$ and this completes the proof of the Lemma.

LEMMA 2. Let H_0 correspond to V_0 as in Theorem I of the preceding section. Then H_0 is maximal symmetric and $-H_0$ corresponds to V_0^{-1}.

Since $\mathfrak{R}_{V_0-1}$ is dense, there is a symmetric H whose Cayley transform is V_0. Corollary 3 of §1, above shows that H_0 is maximal symmetric. Lemma 1(d) and Lemma 4 of §1 show that $-H_0$ corresponds to V_0^{-1}

LEMMA 3. Let V be maximal isometric, i.e. either $\mathfrak{D}_V = \mathfrak{H}$ or $\mathfrak{R}_V = \mathfrak{H}$. Then we can find a finite or infinite set of manifolds $\mathfrak{M}_0$, $\mathfrak{M}_1$, $\mathfrak{M}_2$, ... with corresponding projections E_0, E_1, E_2, ... having the properties: 1. $E_\alpha V = VE_\alpha = E_\alpha VE_\alpha$ $1 = \Sigma_{\alpha=0} E_\alpha$, $V = \Sigma_{\alpha=0} E_\alpha VE_\alpha$. 2. If $\mathfrak{M}_0 \neq \{\theta\}$, then E_0VE_0 when considered on $\mathfrak{M}_0$ alone is unitary. 3. If $\mathfrak{M}_0 \neq \mathfrak{H}$ and (a): $\mathfrak{D}_V = \mathfrak{H}$, then for $\mathfrak{M}_\alpha$, $\alpha \geq 1$, we can find an orthonormal set $\phi_{\alpha,0}$, $\phi_{\alpha,1}$, ... such that $\mathfrak{M}(\{\phi_{\alpha,0}, \phi_{\alpha,1}, \ldots\}) = \mathfrak{M}_\alpha$ and $E_\alpha VE_\alpha(\Sigma_{\beta=0} a_\beta \phi_{\alpha,\beta}) = \Sigma_{\beta=0} a_\beta \phi_{\alpha,\beta+1}$, that is V_α on $\mathfrak{M}_\alpha$ is precisely analogous to V_0 above. If $\mathfrak{M}_0 \neq \mathfrak{H}$ and (b): $\mathfrak{R}_V = \mathfrak{H}$, then for $\mathfrak{M}_\alpha$, $\alpha \geq 1$, we can find an orthonormal set $\phi_{\alpha,0}, \phi_{\alpha,1}$, ... such that $\mathfrak{M}(\{\phi_{\alpha,0}, \phi_{\alpha,1}, \ldots\}) = \mathfrak{M}_\alpha$, $E_\alpha VE_\alpha$ is defined for $\mathfrak{M}(\{\phi_{\alpha,1}, \phi_{\alpha,2}, \ldots\})$ and $E_\alpha VE_\alpha(\Sigma_{\beta=1} a_\beta \phi_{\alpha,\beta}) = \Sigma_{\beta=1} a_\beta \phi_{\alpha,\beta-1}$.

We notice first that if V is such that $\mathfrak{R}_V = \mathfrak{H}$ then $\mathfrak{D}_{V^{-1}} = \mathfrak{H}$. Furthermore one can readily show that 1, 2, and 3(a) for V^{-1} imply 1, 2, and 3(b) for V. Thus we will consider only the case in which $\mathfrak{D}_V = \mathfrak{H}$ since the other case is a consequence of this result.

Suppose then that $\mathfrak{D}_V = \mathfrak{H}$. If also $\mathfrak{R}_V = \mathfrak{H}$ then V is unitary and we let $\mathfrak{M}_0 = \mathfrak{H}$. $\mathfrak{M}_\alpha$ is undefined for $\alpha \geq 1$. We must still consider the case in which $\mathfrak{R}_V \neq \mathfrak{H}$. Let $\mathfrak{N} = \mathfrak{R}_V^\perp \neq \{\theta\}$. We can find an orthonormal set ψ_1, ψ_2, ... such that $\mathfrak{N} = \mathfrak{M}(\{\psi_1, \psi_2, \ldots\})$. Cf. Theorem XI of §6, Chapter II. We note also that V is partially isometric, with initial set $\mathfrak{H}$ and final set $\mathfrak{R}_V = \mathfrak{N}^\perp$. Thus if F is the projection on $\mathfrak{N}$, $V^* = V^{-1}(1-F)$ (Cf. Definition 1 and Lemma 1 of §3, Chapter VI).

Let $\phi_{\alpha,\beta} = V^\beta \psi_\alpha$ for $\beta = 0, 1, \ldots$ and every ψ_α. Since $\phi_{\alpha,\beta}$ is in $\mathfrak{R}_V$ for $\beta = 1, 2, \ldots$, we must have $(\phi_{\alpha,\beta}, \psi_\alpha) = 0$ for $\beta = 1, 2, \ldots$. If $\beta \geq 1$ and $\delta \geq 1$, we also have $(\phi_{\alpha,\beta}, \phi_{\gamma,\delta}) = (V^\beta \psi_\alpha, V^\delta \psi_\gamma) = (V^{\beta-1}\psi_\alpha, V^{\delta-1}\psi_\gamma) = (\phi_{\alpha,\beta-1}, \phi_{\gamma,\delta-1})$. This implies that if $\beta > \delta$, $(\phi_{\alpha,\beta}, \phi_{\gamma,\delta}) = (\phi_{\alpha,\beta-\delta}, \phi_{\gamma,0}) = (\phi_{\alpha,\beta-\delta}, \psi_\gamma) = 0$, while if $\beta = \delta$, $(\phi_{\alpha,\beta}, \phi_{\gamma,\delta}) = (\phi_{\alpha,0}, \phi_{\gamma,0}) = (\psi_\alpha, \psi_\gamma) = \delta_{\alpha,\gamma}$. Thus the $\phi_{\alpha,\beta}$ form an orthonormal set.

Let $\mathfrak{M}_\alpha = \mathfrak{M}(\{\phi_{\alpha,0}, \phi_{\alpha,1}, \ldots\})$ for every α such that ψ_α is defined. From the above, it is readily seen that the $\mathfrak{M}_\alpha$'s are mutually orthogonal. Since $V(\Sigma_{\beta=0}^\infty a_\beta \phi_{\alpha,\beta}) = \Sigma_{\beta=0}^\infty a_\beta \phi_{\alpha,\beta+1}$, $VE_\alpha = E_\alpha VE_\alpha$. Furthermore $V^*(\Sigma_{\beta=0}^\infty a_\beta \phi_{\alpha,\beta}) = V^{-1}(1-F)(\Sigma_{\beta=0}^\infty a_\beta \phi_{\alpha,\beta}) = V^{-1}\Sigma_{\beta=1}^\infty a_\beta \phi_{\alpha,\beta} = \Sigma_{\beta=1}^\infty a_\beta \phi_{\alpha,\beta-1}$. Thus $V^*E_\alpha = E_\alpha V^*E_\alpha$. Taking adjoints, we obtain $E_\alpha V = E_\alpha VE_\alpha = VE_\alpha$.

By Lemmas 5 and 7 of §1, Chapter VI, $\Sigma_{\alpha=1}E_\alpha$ is a projection with range $\mathfrak{M}(\mathfrak{M}_1 \cup \mathfrak{M}_2 \cup \ldots) = \mathfrak{M}(\{\phi_{\alpha,\beta}\})$. Since $E_\alpha V = VE_\alpha$, forming sums and if necessary taking limits, we obtain $(\Sigma_{\alpha=1}E_\alpha)V = V(\Sigma_{\alpha=1}E_\alpha)$. Let $E_0 = 1 - \Sigma_{\alpha=1}E_\alpha$. Then $VE_0 = E_0V = E_0VE_0$, and $1 = \Sigma_{\alpha=0}E_\alpha$. Furthermore $V = 1 \cdot V = (\Sigma_{\alpha=0}E_\alpha)V = \Sigma_{\alpha=0}E_\alpha V = \Sigma_{\alpha=0}E_\alpha VE_\alpha$. The properties listed in (1) have now been established.

The range of $1-E_0$ is $\mathfrak{M}(\{\phi_{\alpha,\beta}\})$ and hence includes $\mathfrak{M}(\{\phi_{\alpha,0}\}) = \mathfrak{M}(\{\psi_\alpha\}) = \mathfrak{N} = \mathfrak{R}_V^\perp$. By Lemma 3 of §1, Chapter VI, we see then that $\mathfrak{M}_0$, the range of E_0, is included in $\mathfrak{R}_V$. Hence E_0V has the same range as E_0. Since $E_0V = VE_0$, we see that $\mathfrak{R}_{VE_0} = \mathfrak{M}_0$. Since $E_0V = VE_0 = E_0VE_0$, we see that V regarded only on $\mathfrak{M}_0$ has range $\mathfrak{M}_0$ and hence is unitary with respect to $\mathfrak{M}_0$ if the latter is not simply $\{\theta\}$. (Naturally $\mathfrak{M}_0$ can be finite dimensional, but the above discussion applies in that case also). Thus we have shown (2). (3) (a) is quite obvious under these circumstances.

LEMMA 4. Let V be isometric, with $\mathfrak{R}_{V-1}$ dense and let H be the corresponding symmetric operator. Let E be a projection with range $\mathfrak{M}$, such that $EV = VE$. Let V' be the contraction of V, with domain $\mathfrak{D}_V \cdot \mathfrak{M}$. Then V' is closed isometric and $\mathfrak{R}_{V'-1}$ is dense in $\mathfrak{M}$. Regarded as a transformation within $\mathfrak{M}$, V' has a corresponding symmetric transformation H' which is the contraction of H with domain $\mathfrak{D}_H \cdot \mathfrak{M}$.

Since $\mathfrak{D}_V \cdot \mathfrak{M}$ is additive and closed, V must be additive and closed, since V is continuous. (Cf. the comment preceding Def. 6 of §1, Chapter IV). Since V' is additive and a contraction of V , it must be isometric. (Cf. Definition 2 of §2, Chapter VI).

Now $\mathfrak{R}_{E(V-1)}$ is dense in $\mathfrak{M}$, since $\mathfrak{R}_{V-1}$ is dense in $\mathfrak{H}$.

Also $\mathfrak{R}_{E(V-1)} = \mathfrak{R}_{EV-E} = \mathfrak{R}_{VE-E} = \mathfrak{R}_{(V-1)E} = \mathfrak{R}_{V'-1}$. Thus $\mathfrak{R}_{V'-1}$ is dense in $\mathfrak{M}$.

Let H' denote the symmetric transformation on $\mathfrak{M}$, which corresponds to V' regarded as a transformation on $\mathfrak{M}$ alone. Let f be in the domain of H'. Then correspondingly we will have a ϕ in $\mathfrak{D}_V\cdot\mathfrak{M}$ such that $f = \frac{1}{2}i(V'-1)\phi = \frac{1}{2}i(V-1)\phi$. Hence f is in the domain of H and we also have $Hf = \frac{1}{2}(V+1)\phi = \frac{1}{2}(V'+1)\phi = H'f$. Thus H' is included in the contraction of H with domain $\mathfrak{D}_H\cdot\mathfrak{M}$.

On the other hand if f is in $\mathfrak{D}_H\cdot\mathfrak{M}$, we have for a $\phi \in \mathfrak{D}_V$, $f = \frac{1}{2}i(V-1)\phi = Ef = \frac{1}{2}i(EV-E)\phi = \frac{1}{2}i(VE-E)\phi = \frac{1}{2}i(V-1)E\phi$. Since $(V-1)^{-1}$ exists, (Cf. Lemmas 3 and 4 of the preceding section), we must have $\phi = E\phi$ and $\phi \in \mathfrak{D}_V\cdot\mathfrak{M}$. Thus $f = \frac{1}{2}i(V-1)\phi = \frac{1}{2}i\cdot(V'-1)\phi$ is in the domain of H'. Thus $f \in \mathfrak{D}_H\cdot\mathfrak{M}$ implies $f \in \mathfrak{D}_{H'}$ and H contracted to $\mathfrak{D}_H\cdot\mathfrak{M}$ cannot be a proper extension of H'. This and the result of the preceding paragraph imply the conclusion of the Lemma.

We also have by (1) of Lemma 2 of §1 above, $EH \subset HE$. Thus if $f \in \mathfrak{D}_H$, Ef is in $\mathfrak{D}_H$ and $Ehf = HEf$. If $f \in \mathfrak{D}_H\cdot\mathfrak{M}$, this means $EHf = Hf = H'f$. Combining these statements, we get that for every $f \in \mathfrak{D}_H$, $Ef \in \mathfrak{D}_H\cdot\mathfrak{M}$ and $EH(Ef) = H'Ef$. Furthermore $EHf = HEf$ implies $EHf = E^2Hf = E(EH)f = E(HE)f = EHEf = H'Ef$. Thus we have shown the corollary:

COROLLARY 1. If E, H and H' are as in Lemma 4 above and $f \in \mathfrak{D}_H$, then $Ef \in \mathfrak{D}_H\cdot\mathfrak{M}$ and $EHf = EHEf = H'Ef$.

If H is maximal symmetric, $V_H = V$ is maximal isometric and either $\mathfrak{D}_V = \mathfrak{H}$ or $\mathfrak{R}_V = \mathfrak{H}$. If $\mathfrak{D}_V = \mathfrak{H}$, we have $\mathfrak{M}_0, \mathfrak{M}_1, \ldots$ and $E_0, E_1, \ldots.$ as in Lemma 3 above, and if we apply the Corollary to Lemma 4 above, we get that for every $f \in \mathfrak{D}_H$, $Hf = (\Sigma_{\alpha=0}E_\alpha)\cdot H = \Sigma_{\alpha=0}H^{(\alpha)}E_\alpha f$, where $H^{(\alpha)}$ is H contracted to $\mathfrak{D}_H\cdot\mathfrak{M}_\alpha$ and is considered as a transformation on $\mathfrak{M}_\alpha$. (2) of Lemma 3 above, and Corollary 1 to Theorem I of the preceding section and Lemma 4 above show that $H^{(0)}$ is a self-adjoint transformation if $\mathfrak{M}_0 \neq \{\Theta\}$. On the other hand, $H^{(\alpha)}$ for $\alpha = 1,2,\ldots$, when they are defined are each analogous to H_0 of Lemma 2 above, with respect to the orthonormal set $\phi_{\alpha,0}$, $\phi_{\alpha,1}, \ldots.$

If $\mathfrak{R}_V = \mathfrak{H}$, the situation is similar except that $H^{(\alpha)}$ for $\alpha = 1, 2, \ldots$, if there are any such, are each analogous to $-H_0$ of Lemma 2 above with respect to an orthonormal set $\phi_{\alpha,0}$, $\phi_{\alpha,1}, \ldots$ Thus we may state

THEOREM II. Suppose H is maximal symmetric. Then there exists mutually orthogonal manifolds, $\mathfrak{M}_0$, $\mathfrak{M}_1$, $\mathfrak{M}_2$, with projections, E_0, E_1, ... respectively, such that $\Sigma_{\alpha=0} E_\alpha = 1$ and such that $E_\alpha H \subset HE_\alpha$ for every E_α. Let $H^{(\alpha)}$ denote the contraction of H with domain $\mathfrak{D}_H \cdot \mathfrak{M}^{(\alpha)}$ and which is regarded as a transformation on $\mathfrak{M}_\alpha$. Then for every $f \in \mathfrak{D}_H$, $Hf = \Sigma_{\alpha=0} H^{(\alpha)} E_\alpha f$. Furthermore if $\mathfrak{M}_0 \neq \{\theta\}$, $H^{(0)}$ is self-adjoint. If $\mathfrak{M}_0 \neq \mathfrak{H}$, we have at least one $\mathfrak{M}_\alpha$ for an $\alpha \geqq 1$ and two cases are possible: (a) If V is the Cayley transform of H, then $\mathfrak{D}_V = \mathfrak{H}$ and $H^{(\alpha)}$ for the $\alpha \geqq 1$, is such that there is an orthonormal sequence $\phi_{\alpha,0}, \phi_{\alpha,1}, \ldots$ for which $H^{(\alpha)}$ is analogous to the H_0 of Lemma 2 or (b) $\mathfrak{R}_V = \mathfrak{H}$ and $H^{(\alpha)}$ for the $\alpha \geqq 1$ is such that there is an orthonormal sequence, $\phi_{\alpha,0}, \phi_{\alpha,1}, \ldots$ for which $H^{(\alpha)}$ is analogous to $-H_0$ of Lemma 2.

The converse to this result can also be given.

Let a number of Hilbert spaces, $\mathfrak{H}_0$, $\mathfrak{H}_1$, $\mathfrak{H}_2, \ldots$ be given, and consider a self-adjoint $H^{(0)}$ in $\mathfrak{H}_0$ and realize H_0 as a $H^{(\alpha)}$ in each $\mathfrak{H}_\alpha$ for $\alpha = 1, 2, \ldots$. We may form $\mathfrak{H} = \mathfrak{H}_0 \oplus \mathfrak{H}_1 \oplus \ldots$ as in Definition 1 or 2 of §3, Chapter III. Let $H\{f_0, f_1, \ldots\} = \{H^{(0)}f_0, H^{(1)}f_1, \ldots\}$ when both sequences exist and are in $\mathfrak{H}$. Now if $V^{(\alpha)}$ is the Cayley transform of $H^{(\alpha)}$, we know from Lemma 8 of §2, Chapter VI that $V = V_0 \oplus V_1 \oplus \ldots$ is isometric with $\mathfrak{D}_V = \mathfrak{H}$. One can also readily show that $V\{f_0, f_1, \ldots\} = \{V^{(0)}f_0, V^{(1)}f_1, \ldots\}$, that $\mathfrak{R}_{V-1}$ is dense, and that H corresponds to V as in Theorem I of §1 above. Thus H is maximal symmetric.

If we take $H^{(\alpha)}$ as a realization of $-H_0$, we have $\mathfrak{R}_V = \mathfrak{H}$ and H is again maximal symmetric. Thus we have:

COROLLARY 1. If H is constructed as in the above, H is maximal symmetric.

CHAPTER XI

REFERENCES TO FURTHER DEVELOPMENTS

Our main purpose in this Chapter is to give references in a number of topics for further reading. We will also give a brief heuristic introduction to each topic.

Our references will be numbered as they are introduced. Two essential references are the following:

(1) M. H. Stone. "Linear Transformations in Hilbert Space". Amer. Math. Soc. Colloquium Publications. Vol. XV, New York, N.Y. (1932).

(2) J. v. Neumann. Princeton Lecture Notes, for the years, 1933 -34, 1934 - 35.

§1

In the footnote to Definition 1 of Chapter VII, §2, we indicated two different kinds of resolutions of the identity. Other types are possible; for instance: Let $E_2(\lambda)$ denote the example (b) in this footnote and let $\phi(\lambda)$ denote a monotonically increasing continuous function with $\phi(\lambda) = 0$ for $\lambda \leqq 0$, $\phi(\lambda) = 1$ for $\lambda \geqq 1$, and $\phi(\lambda)$ has variation zero on the complement of a closed set of measure zero, in the interval $0 \leqq \lambda \leqq 1$. Then $F(\lambda) = E_2(\phi(\lambda))$ offers another example of a resolution of the identity.

Furthermore combinations of these cases occur. For instance: Let $E_1(\lambda)$ denote the example (a) of the footnote referred to in the previous paragraph and $E_2(\lambda)$ denote the example (b). Suppose that these are realized in the two spaces $\mathfrak{H}_1$ and $\mathfrak{H}_2$ respectively. Then in $\mathfrak{H}_1 \oplus \mathfrak{H}_2$, we may consider the projections defined by the equation $G(\lambda)\{f_1,f_2\} = \{E_1(\lambda)f_1,E_2(\lambda)f_2\}$. This again forms a resolution of the identity.

Since $E_1(\lambda)$ varies only on a discrete set of points, $E_1(\lambda)$ is said to have pure point spectra. $E_2(\lambda)$ and $F(\lambda)$ have what is termed continuous spectra, while $G(\lambda)$ has a mixed spectra.

However all the examples cited above have one property in common. Let $\mathfrak{U}_f$ denote the set of elements g for which there is a function $\phi(\lambda)$ such that $g = \int_{-\infty}^{\infty}\phi(\lambda)dE(\lambda)f$. Now in each case, there is an f such that $\mathfrak{M}_f = [\mathfrak{U}_f] = \mathfrak{H}$. Therefore, these resolutions are said to have simple spectra.

But this is not the case in certain other examples. Suppose we consider $\mathfrak{H}_1 \oplus \ldots \oplus \mathfrak{H}_n$ as in definition 1 of §3, Chapter III, and let $E_2(\lambda)$ be realized in each $\mathfrak{H}_\alpha$, $\alpha = 1, \ldots, n$. Let $G_2(\lambda)\{f_1, \ldots, f_n\} = \{E_2(\lambda)f_1, \ldots, E_2(\lambda)f_n\}$. Then the least number m such that there are m elements, $f_1, \ldots, f_m$ such that $\mathfrak{M}_{f_1} \oplus \ldots \oplus \mathfrak{M}_{f_m} = \mathfrak{H}$ is n and thus $G(\lambda)$ is said to have n'tuple spectra.

A complete analysis of these possibilities is given in Stone's treatise, reference (1) above in the following places: Chapter IV, §2; Chapter V, §5; Chapter VI, §1; Chapter VII. (It is believed that these can be read in this order, by a person familiar with the material of this book).

§2

The Operational Calculus

If $p(x)$ is a polynomial and H is self-adjoint we can define $p(H) = a_nH^n + a_{n-1}H^{n-1} + \ldots + a_0 = \int_{-\infty}^{\infty} p(\lambda)dE(\lambda)$ where $E(\lambda)$ is the resolution of the identity corresponding to H. (Cf. Theorem II of §2, Chapter IX). If $\phi(x)$ is continuous, we can define $\phi(H) = \int_{-\infty}^{\infty} \phi(\lambda)dE(\lambda)$. (Cf. Theorem II of Chapter VII, §3). Lemma 4 of §3, Chapter VII and Lemma 5 of §2, Chapter VII show the connection between the properties of these operators and the corresponding properties of the function $\phi(x)$ itself.

We have considered these only for continuous $\phi(x)$. However the equation

$$(Hf,g) = \int_{-\infty}^{\infty} \phi(\lambda)d(E(\lambda)f,g)$$

offers certain possibilities for generalizations. (Cf. Lemma 2 of §3, Chapter VII). For f fixed, this determines the conjugate of an additive functional of g. When the functional is bounded, there corresponds to this functional an element Hf. (Cf. Theorem IV of §4, Chapter II). Thus far we have considered only the possibility of a Riemann-Stieltjes integral. However, if we consider Radon-Stieltjes integrals, we can define the above integral expression for a wider class of functions $\phi(x)$. This is done in Reference (1) in Chapter VI.

For bounded operators these questions are considered from a different point of view in

(3). J. v. Neumann. "Uber Funktionen von Funktionaloperatoren". Annals of Math. Vol. 32, pp 191-226 (1931).

A direct generalization is given in the following two papers of F. Maeda. Maeda does not interpose the numerical integral.

(4). F. Maeda. "Theory of Vector Valued Set Functions". Jour. of Soc. of the Hiroshima Univ. Vol. 4. pp. 57-91, and pp 141-160.

§3
Commutativity and Normal Operators

Definition 1 of §4, Chapter VII applied only to the case in which one of the operators is self-adjoint. For linear operators, an obvious extension is possible but for unbounded operators, certain difficulties in the domains appear. In the more general case, the notion of commutativity has been discussed from a number of points of view. For instance, if A is linear, we may define commutativity by the inclusions, $AB \subset BA$, $A^*B \subset BA^*$. This is discussed in Chapter 14 of (2) above and also in

(5). J. v. Neumann. "Zur Algebra der Funktionaloperatoren". Math. Annalen. B. 102, pp. 370-427. (1929).

In this connection, we would also like to refer the reader to §1 of Chapter VIII of the reference (1).

A similar situation holds with respect to normal operators in the general unbounded case. Various definitions are given in (5), (p. 406), (1) Definition 8.3;

(6). J. v. Neumann. "On Normal Operators". Proc. of the Nat. Acad. of Sc. Vol. 21, pp. 366 - 369. (1935).

(7). K. Kodaira. "On Some Fundamental Theorems in the Theory of Operators in Hilbert Space". Proc. of the Imp. Acad., Tokyo. Vol. 15, pp. 207 - 210. (1939). (It is this definition which we have used.)

These definitions are all equivalent, since they can be shown to be equivalent to A having the integral representation $A = \int_{S_0} Z dE_1(P)$ of Theorem III of §3, Chapter IX.

The theory of normal operators can be developed much further. We have an operational calculus for the functions of a single normal operator. (Cf. (1), Chapter VIII, §3).

§4

Symmetric and Self-adjoint Operators

There are a number of topics in the study of symmetric operators, which we haven't discussed. We refer the reader to the matters discussed in §2 and §3 of (1), Chapter IX. These deal with the abstract significance of "realness" as applied to operators and also the possibility of approximating symmetric operators by bounded symmetric operators.

Another development having practical significance is the result given in:

(8). K. Friedrichs. "Spektraltheorie Halbbeschränkten Operatoren". Math. Ann. B. 109, pp. 465 - 487. (1934).

This paper describes a general method for obtaining a self-adjoint extension of a symmetric operator, which is bounded below, i.e. $C_- > -\infty$.

It has also been shown that one can construct two symmetric operators H_1 and H_2 so that their domains have only θ in common. This significant result is given as Satz 15 in:

(9). J. v. Neumann. "Zur Theorie der unbeschränkten Matrizen". Jour. f. reine u. angewandte Math. B. 161, pp. 208 - 236. (1929).

§5

Infinite Matrices

If T is a c.a.d.d. transformation, we can find a complete orthonormal set $S, \phi_1, \phi_2, \ldots$ in $\mathfrak{D}_T$ (Cf. Theorem X of §6,

Chapter II). We have for each such orthonormal set S an infinite matrix $(a_{\alpha,\beta})$ with $a_{\alpha,\beta} = (T\phi_\alpha, \phi_\beta)$. Furthermore, the infinite matrix and the orthonormal set, will determine for each α, the value of $T\phi_\alpha = \Sigma_\beta a_{\alpha,\beta}\phi_\beta$ if $\Sigma_\alpha |a_{\alpha,\beta}|^2 < \infty$.

However the following possibility may occur. Let $\overline{S}$ denote the set of pairs $\{\phi_\alpha, T\phi_\alpha\}$ in $\mathfrak{H} \oplus \mathfrak{H}$. For a given S, T determines the matrix $(a_{\alpha,\beta})$ and thus $\overline{S}$, nevertheless $\mathfrak{M}(\overline{S})$ may be a proper subset of the graph of T. An example can easily be given. Let $\mathfrak{D}$ consist of all functions f in $\mathfrak{L}_2$ in the form $a_0 + \int_0^x g(\xi)d\xi$ where $g(\xi)$ is also in $\mathfrak{L}_2$. Let Tf $= g$. One can readily see that T is c.a.d.d. The complete orthonormal set $\{\exp(i2\pi nx)\}, n = 0, \pm 1, \pm 2, \ldots$ is in $\mathfrak{D}$ but the pair $\{e^x - e^{1-x}, e^x + e^{1-x}\}$, in the graph of T is orthogonal to the corresponding $\overline{S}$.

This cannot happen for bounded operators T and for these a satisfactory matrix theory exists. The reader is referred to reference (9) for a more general discussion and to (1) Chapter III, §1, which also contains an interesting historical comment.

§6

Operators of Finite Norm

A specialized but nevertheless interesting class of operators is the set of T's whose matrices possess the property that $\Sigma_{\alpha,\beta} |a_{\alpha,\beta}|^2 < \infty$. These are said to be of finite norm and are discussed in (1) Chapter II, §3, Definition 2:15 et seq., Chapter III, §2, and Chapter V, Theorem 5.14.

§7

Stone's Theorem

If we have a family of unitary operators, $\mathfrak{U}(\lambda)$ defined for $-\infty < \lambda < \infty$ having the properties that $\mathfrak{U}(\lambda_1)\cdot\mathfrak{U}(\lambda_2) = \mathfrak{U}(\lambda_1+\lambda_2)$ and that $(\mathfrak{U}(\lambda)f, g)$ is a continuous function of λ for every f and g, then there exists a self-adjoint H such that $\mathfrak{U}(\lambda) = \exp(i\lambda H)$. This result, which has many important applications, is due to Stone:

(10). M. H. Stone. "On One Parameter Unitary Groups in Hilbert Space". Annals of Math. Vol. 33, pp. 643-648. (1932).

J. v. Neumann pointed out that we may replace the condition of continuity for $(\mathfrak{U}(\lambda)f,g)$ by measurability.

(11). J. v. Neumann. "Uber einen Satz von Herrn. M. H. Stone". Annals of Math. Vol. 33, pp. 567 - 573. (1932).

§8

"Rings of Operators"

We have been concerned up to now with matters which depend on the structure of a single operator, indeed on the structure of a single normal operator. The structure of certain sets of transformations is also interesting and has been studied recently.

These investigations have been concerned with the infinite dimensional equivalent of semi-simple matrix algebras. These sets are also the generalization of group algebras and are of interest because of Haar's result that any locally compact separable topological group can be represented as a set of unitary transformations in Hilbert space.

(12). A. Haar. "Der Massbegriff in der Theorie der kontinuierlichen Gruppen". Annals of Math. Vol. 34, pp. 147 - 169. (1933).

In reference (5) above, J. v. Neumann introduced the notion of ring of operators. A set of linear operators, M, is said to be a ring of operators if $A \in M$ and $B \in M$ imply aA, A^*, $A+B$ and $AB \in M$ and if, furthermore, M is closed in a certain topology. This topology insures that the limit of a weakly convergent sequence of operators of M, is in M. Whether closure in this topology is equivalent to this property, if M has the algebraic properties of a ring, is not known. If M does not have these algebraic properties, this topology is not equivalent to sequential closure.

This topology is desired, however, since rings closed under this topology have the following very interesting property. Let M' denote the set of linear operators which commute with all $A \in M$. Then if M is a ring containing 1, then $(M')' = M$. This result is proven in (5) in a slightly more general form.

Also important is the result proven in (5) and (3), that if M is abelian, i.e. $M' \subset M'$, then there exists a self-adjoint H, such that every transformation in M is a function of H.

In an unpublished work, Professor von Neumann has also established what might be called the resolution of a ring with respect to its center, $M \cdot M'$. $M \cdot M'$ is abelian. Suppose its center contained only a finite number of mutually orthogonal projections $E_1, \ldots, E_n$ and suppose that each E_α is minimal in $M \cdot M'$. If M contains 1, $1 = \Sigma_\alpha E_\alpha$ and if $A \in M$, $A = (\Sigma_\alpha E_\alpha)A = \Sigma_\alpha E_\alpha A E_\alpha$ since $E_\alpha \in M \cdot M'$. If we consider the $E_\alpha A E_\alpha$ on $\mathfrak{M}_\alpha$, the range of E_α, then these transformations form a ring of operators on $\mathfrak{M}_\alpha$ whose center is simply $\{a \cdot 1\}$.

A ring for which $M \cdot M' = \{a \cdot 1\}$ is called a factor. Thus M can be called the $\oplus$ sum of factors. In the general case, the possibility of a continuous spectrum for the H which determines $M \cdot M'$ offers difficulty, but Prof. v. Neumann's result is that for a suitably generalized definition of $\oplus$ sum, every ring M is the $\oplus$ sum of factors.

Thus the analysis of rings of operators in general can be referred to in the study of factors. From the corresponding result in the finite dimensional cases, one would suspect that a factor must be isomorphic to the set of all linear operators on Hilbert space or on a finite dimensional unitary space.

But this is not the case in general as is shown in:

(13). F. J. Murray and J. v. Neumann. "On rings of operators". Annals of Math. Vol. 37, pp. 116 - 229. (1936).

Here one considers a relative dimension function $D_M(E)$ defined for the projections in M and having the properties that $D_M(E_1) \leq D_M(E_2)$ if and only if there exists a partially isometric W in M, whose initial set is the range of E_1 and whose final set is included in the range of E_2. When $D_M(E_1) = D_M(E_2)$, W can be chosen so that the final set is the range of E_2.

If M is a factor isomorphic to all the operators on a Hilbert space, $D_M(E)$ takes on the values $0, 1, 2, \ldots, \infty$ for various E's in M and only these values. A similar result holds if M is isomorphic to the operators on an n-dimensional space. These cases are called respectively I_∞ and I_n.

There are, however, essentially, three other possibilities for the range of $D_M(E)$; case II_1 in which $D_M(E)$ assumes all values α such that $0 \leqq \alpha \leqq 1$, II_∞ in which $0 \leqq \alpha \leqq \infty$ and III_∞ where $\alpha = 0$ or ∞.

Examples of factors in case II_1 and II_∞ were given in (13). An example of a III_∞ was first given in

(14). J. v. Neumann. "On rings of operators III". Annals of Math. Vol. 41, pp. 94 - 161. (1940).

Let M be a factor in a case II_1 and $H = \int_{-C}^{C} \lambda dE(\lambda)$ be a self-adjoint operator in M. The expression

$$T_\Lambda(H) = \int_{-C}^{C} \lambda dD_M(E(\lambda))$$

was defined in (13) and shown to have a number of the properties of a trace in the finite dimensional case. The property, $T_\Lambda(H_1+H_2) = T_\Lambda(H_1)+T_\Lambda(H_2)$ was established in

(15). F. J. Murray and J. v. Neumann. "On rings of operators II". Trans. of the Amer. Math. Soc. Vol. 41, pp. 208 - 248. (1936).

For a fixed $\alpha = 1, 2, \ldots, \infty$, all M in case I_α are isomorphic. However, in a forthcoming joint paper of Professor von Neumann and the writer it will be shown that not all II_1's are isomorphic.

The following paper is also of interest in connection with rings of operators:

(16). J. v. Neumann. "On infinite direct products". Compositio Math. Vol. 6, pp. 1 - 77. (1938).

An application of the Rings of Operators theory to a new development is given in:

(17). F. J. Murray. "Bilinear transformations in Hilbert space." Trans. of the Amer. Math. Soc. Vol. 45, pp. 474 - 507. (1939).

CHAPTER XII

REFERENCES TO APPLICATIONS

In this Chapter, we give a brief list of references to the applications of the theory which we have discussed.

§1

We wish again to refer the reader to reference (1) of the preceding Chapter, in particular to Chapters III and X. There the following topics are discussed: Integral Operators (including the Fourier transforms), Differential Operators, Operators corresponding to flows (briefly), Jacobi matrices and Moment problems.

In connection with integral operators, mention should also be made of:

(18). J. v. Neumann. "Charakterisierung des Spektrums eine Integraloperators". Actualites Sci. et Industrielles, 229, Herman et Cie, Paris. (1935).

This paper is concerned with self-adjoint operators, which may be represented as integral operators. It is characteristic of these operators that there exists no $\epsilon > 0$ such that $E(\epsilon)-E(-\epsilon)$ has a finite dimensional range.

§2

Differential Operators

The nature of the domain of differential operators and the possibility of symmetric extensions have been investigated from a number of viewpoints. Perhaps the most natural continuation of the work of von Neumann and Stone is to be found in the papers,

(19). I. Halperin. "Closures and adjoints of linear differential operators". Annals of Math. Vol. 38, pp. 880 - 919, (1937).

(20). J. W. Calkins. "Abstract symmetric boundary conditions." Trans. of the Amer. Math. Soc. Vol. 45, pp. 369-442. (1939).

The notions of reference (8) of the preceding Chapter have been extended and applied to differential operators in:

(21). K. Friedrichs. "Spektraltheories halbbeschränkten Operatoren. Zweiter Teil". Math. Annalen. B. 109. pp. 685 - 713, (1934).

(22). K. Friedrichs. "Uber die ausgezeichnete Randbedingungen in der Spektraltheorie der halbbeschränkten gewöhnlichen Differentialoperatoren zweiter Ordnung". Math. Annalen B. 112, pp. 1 - 23, (1935).

(23). K. Friedrichs. "On differential operators in Hilbert space". Amer. Jour. of Math. Vol. LXI, pp. 523 - 544, (1939).

Another paper dealing with differential operators is:

(24). F. J. Murray. "Linear Transformations between Hilbert spaces". Trans. of the Amer. Math. Soc. Vol. 37, pp. 301 - 338. (1935).

§3

The very important applications of the notions of linear transformations in Hilbert space to quantum mechanics are explained in:

(25). J. v. Neumann. "Mathematische Grundlagen der Quantenmechanik". J. Springer, Berlin. (1932).

§4

The connection between the theory of Operators and the usual Hamiltonian mechanics was pointed out in:

(26). B. O. Koopman. "Hamiltonian systems and transformations in Hilbert space". Proc. of the Nat. Acad. of Sci. Vol. 17, pp. 315 - 318. (1931).

This connection is based on Stone's Theorem (Cf. reference (10) and (11), and has led to many interesting and important results. These are given in:

(27). B. O. Koopman and J. vonNeumann. "Dynamical systems of Continuous Spectra." Proc. of the National Acad. of Sci., Vol. 18, pp. 255-263. (1932).

(28). J. vonNeumann. "Proof of the quasi-ergodic hypothesis." Proc. of the National Acad. of Sci., Vol. 18, pp. 70-82 (1932).

(29). J. vonNeumann. "Zur Operatorenmethode in der klassichen Mechanik." Annals of Math., Vol. 33, pp. 587-642 (1932).

REFERENCES

References to the literature are given on pages 3, 7, 15 and in Chapters XI and XII.

The following indices refer only to Chapters I to X.

INDEX OF SYMBOLS

Non-litteral Symbols

Litteral Symbols

INDEX OF TERMS

Adjectives are given in connection with the associated noun.

Lightning Source UK Ltd.
Milton Keynes UK
UKHW040620200522
403285UK00001B/83

9 780691 095691